Shade

Shade

The Promise of a Forgotten Natural Resource

Sam Bloch

RANDOM HOUSE · NEW YORK

Random House
An imprint and division of Penguin Random House LLC
1745 Broadway, New York, NY 10019
randomhousebooks.com
penguinrandomhouse.com

Portions of this book were previously published in *Places Journal* and *Slate*.
Art credits are located on page 297.

Hardcover ISBN 978-0-593-24276-6
Ebook ISBN 978-0-593-24277-3

Printed in the United States of America on acid-free paper

9 8 7 6 5 4 3 2 1

First Edition

BOOK TEAM:
Managing editor: Rebecca Berlant • Production manager: Richard Elman
Copyeditor: Maureen Clark • Proofreaders: Marisa Crumb, Ethan Campbell

Title-page and Part title-page art: Sue Tansirimas/Adobe Stock

Book design by Edwin A. Vazquez

The authorized representative in the EU for product safety and compliance is
Penguin Random House Ireland,
Morrison Chambers, 32 Nassau Street,
Dublin D02 YH68, Ireland.
https://eu-contact.penguin.ie

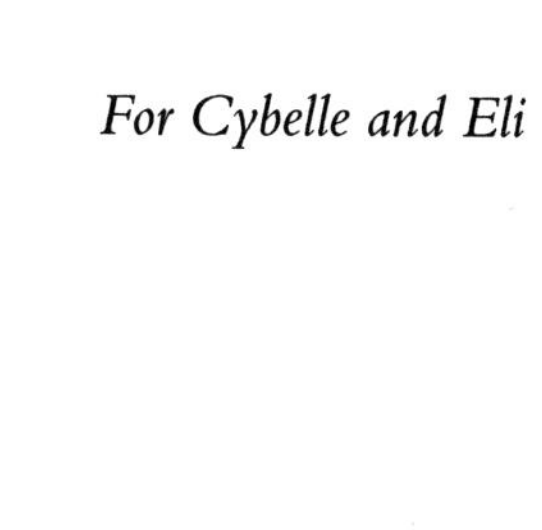

For Cybelle and Eli

Contents

Introduction

As the sun rises in Los Angeles, a dozen passengers wait for a downtown bus in front of Tony's Barber Shop. On this barren stretch of Figueroa Street near the Pasadena Freeway, they stand one behind the other, still and quiet in the shadow of the person in front of them. It's going to be another scorching summer day, and across the city, riders like these are hiding from the sun behind road signs, telephone poles, and whatever meager shelter they can contrive.

For years, Cypress Park's bus riders suffered the pulsing star's enervating rays on a pitiless street. City officials dismissed their pleas for a bus shelter with a roof and a bench, claiming the sidewalk was too small to fit one in. Puny trees dug out of the dirt withered without water or care. All the while, the sun beat down on a hundred-foot span of asphalt and concrete, heating the hard ground and warming the air.

The conditions might have been more tolerable if the bus came more often, but thirty-minute headways are the norm on some routes in this part of town. An unshaded wait is enough to ruin a typical commuter's day, and for others—the elderly, people in poor health, and those with physical disabilities—it's the beginning of something more serious. Doctors have medical terms for the illnesses that result from heat exposure, but Tony Cornejo, the barbershop's curmudgeonly proprietor, had his own way of describing what happens to bus riders on Figueroa: "They're burning themselves out there."

The old-school Cornejo had a soft spot for them. In 2014, a leafy banana tree and a metal pole appeared in an unsealed patch of sidewalk near the curb. He swore he didn't put them there, but he admitted to hooking a gray canvas to the pole and stretching it over the sidewalk to create a public canopy. "We needed something like that for the ladies and children," who otherwise stood in his shop to get out of the sun, he explained.

To make the canopy more inviting, Cornejo borrowed some wooden crates from the supermarket next door. He dragged them into the shade and nailed them together to make two parallel benches. Now the families had somewhere to sit while they waited. "They loved it," Cornejo said, beaming. Initially, he thought bus riders would be the beneficiaries of his street shelter, but over time it seemed the whole neighborhood came to sit in his shade. Men thumbed through magazines. A Metro bus driver scrolled on his phone before his shift. A frail old-timer waited patiently for his ride, hands folded over a quad cane. Even a local can collector came by, dragging his shopping cart to the streetside oasis. When wind and rain ruined the canvas, Cornejo replaced it with a sturdier wood roof. He was just taking care of the sidewalk, he told me, no different from sweeping it clean in the morning.

But in 2015, the city's sidewalk inspectors heard about the shelter, and ordered Cornejo to take it down. According to city rules, sidewalks have to be safe and secure, and apparently, it posed a threat. Sidewalks also have to be accessible to people with disabilities, and according to the authorities, the seat in the shade made Figueroa Street more hostile to the man with the cane. As temperatures rise, the definition of public safety could include protection from the sun. But L.A.'s sidewalk codes have in fact become more punitive and quashed grassroots improvements in underserved neighborhoods like Cypress Park. Cornejo didn't want any trouble. He asked a customer to help him break down the lumber and return the sidewalk to its previous condition, just in time for L.A.'s worst heat wave in twenty-five years.

We need to manage heat to live. We have an effective and democratic way of doing it. And yet, as the planet warms, the powers that be reject it. Why?

Every year, heat takes more lives than floods, hurricanes, and tornadoes combined, but the fatalities often go unremarked, perhaps because the danger is invisible. There's no twister that uproots a neighborhood, no flood that sucks it underwater, nor billions of dollars in property damage. Instead, heat's imprint is seen in empty streets, work slowdowns, cognitive decline, and hospital bills. When autumn comes, and temperatures come back down to earth, it leaves no visible trace.

Scientists are certain the earth is getting hotter. L.A. summers, and those just about everywhere else on the planet, are already two to three weeks longer than they were in the 1950s. By the end of the century, the warm season in the United States could last six months, and extreme temperatures could force us to spend much of it indoors. Supercharged heat waves will settle over cities for weeks at a time and cause epidemics of death. The survivors will suffer heart attacks, kidney disease, and brain damage. What we now call winter will be a brief, two-month interregnum that mostly feels like spring.

What are we going to do? Your first thought is probably a reasonable one: Stop burning coal, oil, and natural gas, the fossil fuels that release carbon dioxide and other heat-trapping gases into the atmosphere. You're right. It's the best way to stop this madness from getting any worse. But those gases aren't going anywhere. Carbon dioxide takes three hundred to one thousand years to leave the atmosphere and return home to the deepest sediments of the earth. Even if every single power source becomes a renewable one and we stop emitting carbon, the planet's surface won't start cooling. The temperature will continue to rise for a few years before gradually tapering and leveling off. And then we will be living on an artificially heated planet for a very long time. It will take "many, many centuries," NASA estimates, to end the global greenhouse effect, perhaps no earlier than the year 3000, if not much later. It is a sobering truth that cutting emissions isn't enough. We also need to begin a new life on a new earth.

What if the key to that new life is as old as civilization itself?

Shade is what public transit riders want. They deserve to wait for the bus with a basic level of comfort and dignity. Shade is what outdoor workers need. Farmhands and construction crews are most at risk from heat's fatal impacts, and amid the hottest summers in human history, their advocates have called on legislators to ensure the right to water, rest, and shade.

Shade is what environmentalists desire. Across the United States, a network of nonprofits have fanned out to plant trees in neighborhoods that need them most, arguing that in the climate change era, green and shady streets are necessities for everyone. Their work is bolstered by powerful research that traces today's urban heat disparities back to redlining, the racist lending policies that denied Black Americans and other people of color the capital they needed to buy a home and invest in their neighborhoods.

And shade is what heat experts suggest. They used to make their case in the pages of obscure scientific journals, but now that extreme heat is front-page news, they are taking their message to Americans on Capitol Hill, cable TV, and chart-topping podcasts. "We all know that cities are cooler when we have shade, but we're not really planning for it," said V. Kelly Turner, an urban planning and geography professor at the University of California, Los Angeles. "In the future, that's something that cities are going to need to do, is intentionally think about, what does shade infrastructure look like?"

Turner believes shade could be America's next long-term investment in public health. What safe drinking water and clean air were to the twentieth century, shade could be to the climate-changed twenty-first. Scientific models bear her out. If we can get emissions under control and put the planet on the path to moderate warming, then by 2050 getting out of the sun could be the difference between forty-one days of unsafe heat in humid New Orleans and none at all. Between 102 days of lethal temperatures in bone-dry Phoenix and nine days of danger. Between the oppressive conditions of an unshaded bus stop that push a vulnerable person to the brink and a safe place to hang outside.

So if we're going to need more shade in the future, what's the best kind?

If you answered that question in your head, your answer was probably trees—and for good reason. The human species evolved in forests, and some of our oldest myths and stories unfold in their speckled light. Hippocrates taught medicine under a plane tree, and Ovid found bittersweet beauty in the laurel's leaves. The Mesopotamian goddess Inanna slept under a miraculous poplar whose shadow never moved, Buddha found enlightenment by meditating under a ficus tree, and Christian and Muslim heavens alike are cooled in their perpetual shade.

Here on Earth, tree shade is where public space was born and civic identities were forged. In Accra, Khartoum, and Kinshasa, leafy canopies are the roofs of outdoor rooms where barbers cut hair, hawkers haggle with customers, and tea is served. Shade takes on great importance in the remote desert villages of Burkina Faso, where there's no air-conditioning and scant indoor lighting. In the daytime, people gather under neems and enormous baobabs that burst from scrubby red soils to become de facto town halls. As the Burkinabe architect Francis Kéré has observed, the shade is their court and the roots their benches.

In tropical climates, people naturally prefer to confer, hustle, and gossip out of the sun's permanent glare. Artists and landscape architects return from Mexico and Cuba thunderstruck by shade's ruthless dictation of social choreography, how the shadows of buildings and flagpoles accidentally determine where people meet. Environmental researchers in Marrakech, New Delhi, and Singapore see that people spend far more time in a shady park or temple courtyard than in a sunny one. They linger and relax, and that engenders more interactions, and possibly even stimulates social cohesion. It's true in arid cities, humid regions, and even temperate zones with shorter summers. People want to be in shade. They muse longer, pray more peacefully, and find strength to walk farther.

It's understandable that Americans have forgotten how sweet shade can be. As air-conditioning has become the default method of cooling down, the *shade tree* has disappeared from the lexicon. These days, we appreciate trees for so many other reasons: the wildlife harbored, the fruits flowered, the cleaner air, the soft rustle of foliage, the pastoral beauty, and the carbon sequestered in their trunks. Yet there is still no technology known to man that cools the outdoors as effectively as a tree. These communal parasols are also misting machines that cool the air. It's hard to feel that effect under one or two of them, but get enough trees together and an urban summer can be as fresh as a rural spring, a feat with major implications for energy use and public health.

Trees aren't the only public infrastructure that casts shade. Around L.A., on streets like Figueroa that are too cramped and paved over to support green canopies, the preferred protections aren't arboreal but artificial, like the pop-up tents of *taqueros* and the cheerful rainbow umbrellas of fruit vendors. In Phoenix, a desert city that struggles to nourish an urban forest, sidewalk screens, frilly metal filters, and soaring photovoltaic canopies are more common. These interventions are more impactful than you might expect. Ariane Middel, an Arizona State University urban climate researcher who runs the school's Sensable Heatscapes and Digital Environments (SHaDE) Lab, surveyed students and staff as they strolled through the shadows of solar panels cast on a campus thoroughfare. More than any change to ambient temperature, humidity, or wind, the mere presence of shade was the only significant predictor of outdoor comfort in Tempe.

Shade's effectiveness is a function of physics. It depends on the material properties of the sun-blocking objects that cast it—how they reflect, absorb, and transmit different wavelengths of energy in sunlight. It depends on the intensity of that light and the extent of the shade thrown. (A telephone pole that casts a perfect shadow on your body does nothing to stop the solar heating of the surfaces around you.) And it depends on the biology of the person who receives it. Middel has come as close as anyone to adding up all these factors. She praises humble umbrellas and plastic sails, because their shade feels like taking 30 degrees off the afternoon sun, which is about as good as shade cast by a tree. Ultimately, she finds the city itself offers the most relief in the shadows of arcaded sidewalks and looming skyscrapers. This is an ancient truth that we have largely forgotten, and the rediscovery could change what we think we know about heat.

If shade is as old as the Bible, so is the bias against it. After the original sin of eating the fruit from the tree of knowledge, Adam and Eve hid from God in the verdure to conceal their guilt. When something is dubious or seems criminal, it's *shady*. And when we are offended by someone, we are said to *take umbrage*. That expression might date back to the ancient Romans, who sneered at the shady characters (*umbratici*) who sold their bodies and begged under the porticoes, or the English poet John Milton, who rendered the expulsion from the Garden of Eden as a journey "from shadowy types to truth."

Think of the murky shadows under a bridge or highway overpass or the menacing darkness of a narrow alley. Who would choose to lurk there? The phrase is already ringing in your ears: Those people are shady. Because no good deed is done in the shade, it's synonymous with insult. We all know about throwing shade. The Black women and drag queens who invented the phrase say it's the art of the side-long insult, and a clever way to dress down a rival. As Dorian Corey puts it in the 1990 documentary *Paris Is Burning,* "Shade is, I don't tell you you're ugly, but I don't have to tell you, because you know you're ugly."

Americans are sun-loving people. We think shade is yucky. It's for damp corners and fetid ponds, the habitat of swarming bugs and mosquitoes where bacteria fester and plants die. It's no good for us. The Greek philosopher Onesicritus taught that shade stunted growth, a belief that presaged a modern fixation on the healthiness of sunlight. In the nineteenth and early twentieth centuries, doctors and reformers feared that darkness itself caused the poor health of urban slum dwellers. It was a vector of disease, where contagions bred and spread, and the murkiness additionally encouraged licentiousness and other urban vices. Some literally believed sunlight was the best disinfectant.

Once, pale skin was desirable, the look of someone who never had to bother with menial outdoor labor. But it wasn't long until the sunlight cure came to America and we embraced a new concept of health. Solar codes were written into urban plans, and new materials and technologies allowed architects to design brighter buildings flooded with natural daylight. Now, we're beginning to see how a solar fetish may be maladaptive. In New York, a recent summer saw a throng of neighborhood activists protest the construction of a sixteen-story office tower with signs to "Save Our Light." They did this while huddling in the shadow of another building. As the planet warms, and the sun becomes more foe than friend, perhaps we should consider the dark side of a bright future.

This book emerged from an article I published in 2019. Back then, I reflected upon my experiences in Los Angeles and the struggles of unsung heroes like Tony Cornejo to improve their neighborhoods. Rather than discourage shade, I urged L.A. and other warming cities to mandate it in urban design. At the time, this seemed like a naïve and unrealistic goal. Since then, to my surprise, I have seen shade embraced. It has become a focal point in an emerging debate about protecting vulnerable people from climate change and closing a widening gap of heat exposure. Part of that debate is about public space. Should streets, sidewalks, parks, and schoolyards remain denuded to ease the lives of drivers and cops? Or should they be aggressively shaded to promote better health? Shade can save our cities from becoming uninhabitable wastelands, but only if its value is grasped by the people who lead them.

This book draws from the past to inform the future. Part I is a shadow history of humanity that shows shade's place in the natural world as an elemental shelter. It visits the earliest settlements and grandest cities where shade was a foundation of culture. And it explores how the human species was lured inside, where it tuned out the natural world, with great consequences for the global climate.

Part II is about the here and now. It explores the effects of shade on individual and collective scales, heading to farm fields where workers suffer in the sun. It teases out the causes of shade's unequal provision in cities, showing how urban planning decisions can impact human health during heat waves. It is seldom appreciated that cities, where most of us live, are actually warming faster than the planet as a whole, for reasons that are also anthropogenic—we built them that way. Cities are where the provision of shade could have the greatest health impact. Unfortunately, we don't have much time to lose. Poorer city dwellers are already falling sick and dying because their neighborhoods aren't as shady as wealthier, whiter ones. Inaction today puts future generations at risk. Will we force them indoors to rely on expensive and environmentally destructive AC? Or will we help them live comfortably and sustainably on a hotter planet?

Part III looks to the future, probing how shade could help us adapt to climate change. It tours American, European, African, Asian, and Australian cities where new homes and schools are built sustainably to withstand heat, and where streets are retrofitted to be green. And it floats to the atmosphere, where carbon pollution traps the sun's heat.

We all know we need to stop emitting greenhouse gases, but some of us are frustrated by the glacial pace of progress. What if we could intercept a tiny fraction of solar radiation before it strikes the planet? Climate entrepreneurs proposing to geoengineer the globe with reflective aerosols in the stratosphere wouldn't dapple the light on the sidewalks but dull it in the skies. Just as shade takes many forms on Earth, so too are there many possibilities in the heavens. We are closer than ever to making this dream—or nightmare—into a reality.

As unprecedented heat bears down, we have to find a way to transcend petty debates and see shade as a basic human right. Shade's fall from favor has paralleled and fueled the destruction of the global climate. If we reintegrate shade into our daily lives in small and big ways, the impacts could be thorough and profound—from a subtle return to a life outside, to the slowing of warming and the righting of social wrongs. We have forgotten that shade is a natural resource, and for that reason, it is neither protected nor widely cast. We don't remember what it's done for us, we don't grasp its importance, and we don't appreciate its promise for a better future. Loggers and farmers cut down forests, forcing animals to flee and land to turn fallow. Engineers ignore time-honored methods of keeping out heat, locking us into mechanical cooling systems that fail during blackouts. And urban planners denude shady parks and pave neighborhoods under heat-sucking roads, only to drive us mad by the infernal conditions. But it's not too late to start talking about shade again. So let's begin.

Part I

A Brief History of Shade

Chapter 1

Made in the Shade: Why Shade Is an Evolutionary Imperative

On a balmy afternoon in rural Oregon, a flat, shallow creek riffles gently over gravel. The creek runs straight through a sunny meadow for as far as the eye can see, the centerline of a warm, pastoral scene that draws the gaze from a county road to brown cows roaming on grass, to the foothills studded with bushy trees in the distance. Generations ago, the loggers who cleared this land for lumber worked its water into submission, straightening a muddy, meandering path into an agreeable channel, and dredging a rough bottom into a smooth floor. Today, the creek, part of the headwaters of the John Day River, moves seamlessly through the pasture, wetting the grass in the dry summer months.

Where the creek empties into the river, the water regains its natural, more erratic flow. The path shallows and deepens, and the channel splits and curves around hairy bunchgrasses and thick tufts of willow that push through the banks. The brush grows larger and consumes the stream, crowding it under woody thickets of branches and stems. Farther into the backcountry, the water disappears entirely into wilderness, hiding in dense forests and slicing through steep, rocky ravines. When the sun sets over the hills, the waning light breaks through the leaves to dance on the river's glistening surface and flash the areas under the banks where rushing water has eroded the soil. And down in one of those undercuts, lurking behind the dangling roots of shrubs and brush, a Chinook salmon waits to spawn.

The king of salmon is a heavy, substantial animal, weighing thirty pounds and measuring thirty inches from snout to tail. Two months ago, this torpedo-shaped beast left the frigid waters of the Pacific Ocean to begin a long, arduous journey, swimming hundreds of miles upstream against the high-flowing Columbia River with only its reserves of fat as fuel. The fish ascended thousands of feet to navigate over two dams and traverse a network of open rivers, sunny meadows, and shrubby streams before finally arriving here, at this hidden nook in the forest.

There are countless places along the run where she could have stopped—why here? Even the royals of the aquatic kingdom are someone else's prey, and she likely chose the undercut to hide from predators like circling ospreys. Although salmon stop feeding when they leave the ocean and begin their heroic migration, this section of the river, rich in leaves, branches, and woody debris, is also abundant in insectile food. When the nightly mayfly hatch begins, the surface of the water erupts with the open mouths of trout and other fish inhaling the aquatic bugs. But above all, she chose this home not for what it provides, but for what it shelters her from.

Ninety-three million miles away, a huge and powerful star pulses with waves of energy, known as radiation, that travel through the vacuum of space and the filter of the atmosphere to impinge on the surface of the earth. Sunlight converts to heat when it touches water, rocks, and fish. Most of the time, animals need this warmth, because it activates their muscles and stimulates their metabolisms. But in the summer, when the Northern Hemisphere tilts toward the sun, and the forks and streams of the John Day River receive its beams more directly, the excess of energy is unwanted.

Even tepid water is dangerous to Chinook salmon, and eyes bulging, they can struggle to breathe. They have a preferred thermal habitat, and in the summertime of the Pacific Northwest, they seek a stream that is out of the sun's glare and made in the shade. High above the river, canopies of cottonwoods, aspens, and pines catch the sunbeams, scattering the light and absorbing the energy that would otherwise raise the temperature of the water to an untenable degree. Even more than the temperature of the land that surrounds it, or that of the warm air gusting above it, exposure to the sun's rays determines whether this river can sustain its animal life.

In the fall, when the water cools, the salmon will dart out from the undercut to make her final ascent upstream. She will beat against the tide one more time to dig her nest in the river's gravel bottom and shoot thousands of eggs from beneath her flapping tail, alongside an eager male ready to catch them. And after this last, most important act of her brief and exceptionally challenging life, she will succumb and float downstream, never once glimpsing her eggs as they begin to hatch. Shade, for Chinook salmon, is the setting of the life story. It determines how these creatures feel and behave, and even where they die. Generations rely on it, and in a few years, her fry will return to this same stream to spawn. Provided that the shade, in all its fragility, is still there.

Chinook salmon are ectotherms: organisms that can't control their body temperature internally. Instead, they regulate their temperature behaviorally, by finding and occupying the right environment. We used to call these animals cold-blooded, but that's misleading. An ectotherm that spends too much time in the sun starts to run hot. It can't cool down by sweating, like a human does. Instead, an ectotherm must thermoregulate by hiding behind a tree, burrowing underground, or in the case of the Chinook salmon, immersing itself in a cooler body of water.

Because ectotherms comprise more than 99 percent of animals, it's not much of a stretch to say that all life on Earth needs shade to survive. Think of the mites, mosquitoes, and spiders that swarm in cool corners and scummy grottoes, or the frogs and salamanders that spawn in the dim, moist environs of vernal pools. Butterflies flit through the air, making habitats in shady microclimates under grasses and leaves. For the benefit of those and other pollinators, some humans leave wild and kinky patches on their otherwise tidy lawns, where the ectotherms can rest in microshade.

Then there are the lizards of the American Southwest. Unlike the salmon in the water, these ectotherms inhabit the air, a gaseous fluid. These featherweight animals have thin skin and hardly any body mass for insulation, which means their bodies absorb the heat in this fluid immediately. And although they have evolved to tolerate the desert's high ambient temperatures, they are vulnerable to gain unwanted heat with each scamper across the burning sand and every sunbeam that strikes them through a cloudless sky.

In the cool morning hours, when heat is needed to raise their bodies to an ideal temperature, these animals perch upon a rock to face the rising sun and bask in its rays. Others prefer to flatten against the surface to expose the maximum amount of skin. It may take only a few minutes for the reptiles to reach their preferred internal state and acquire the energy they need to scamper off and hunt for insects and spiders. But by noon—when the sunlight is no longer smudging inefficiently across the atmosphere but pouring upon the earth's surface and any creature dashing across it—the heat can prove overwhelming. This is around the time that lizards begin to burrow under rocks and nestle into thorny bushes to shield themselves from the sun. Although not for long. Shade denies lizards the opportunity to forage, and sometimes, it even saps them of the will to live. Too hot, too cold—for these diminutive beings, neither microclimate is right. To achieve their ideal temperature, they must shuttle back and forth between sun and shade, smoothing the extremes into a more palatable range.

In 2016, the biologist Michael Sears monitored a lounge of lizards as they scampered around research plots in New Mexico's Sevilleta National Wildlife Refuge in this dance of behavioral thermoregulation. Sears observed that lizards marooned under large canopies tended to hunker down, and in the process, risk starvation. But in shade-speckled landscapes, they roamed the terrain, which kept their temperature intact. Like the nook of a riverbank, a craggy manzanita bush may not look like much to a human. To a small and sensitive ectotherm, it can be a pit stop on the road to survival. Sears urges us to preserve the natural patchworks of shade in our own landscapes to give animals a better shot at traversing a world that we have made ever more hot and hostile.

Rocky crevices, sandy burrows, umbrous inlets, and other cool escapes are what ecologists call thermal refuges. Although most of us hardly notice the shadows they cast on the dirt and the edges of streams, the humans who interact most closely with these ecosystems understand that this microshade is essential.

For two centuries, homesteaders, lumberjacks, and farmers cleared the land around the John Day River for their own use. It wasn't their intention to warm the waters of the river basin, but that was the effect all the same. Rising global temperatures and fatal marine heat waves raised the stakes. Now, Oregon's state environmental regulators classify "solar radiation" (the sun's heat) as a stream pollutant that threatens the salmon's dwindling numbers, alongside toxic waste and chemical runoff, and call for shade as the necessary remediator. The local Wasco, Warm Springs, and Paiute tribes, who have long depended on the salmon, are leading the effort by replanting willows and cottonwood along with other native riparian shrubs and trees. Using modern modeling techniques, U.S. Forest Service scientists have confirmed what the regulators and the Confederated Tribes of Warm Springs already know: By restoring the original riverine shade, it is possible to more than offset all the damage wrought by climate change upon the magnificent beasts.

Travel west from the John Day River toward the Oregon coast and you will enter the murky light of a temperate rainforest. Beneath the massive canopies of ancient maple, spruce, and hemlock trees, the damp forest floor is blanketed in mosses, ferns, and shrubs. The ecosystem offers a glimpse of the spectacular biodiversity found in the understories of rainforests around the world. In the Amazon, the Congo basin, and the jungles of Borneo, seedlings sprout in the thick, humid underbellies of hundred-foot-tall trees. Some of these plants are spices and fruits: cinnamon, vanilla, cocoa, and nutmeg, along with jackfruit, durian, mango, and coffee. Others end up in your apartment, as houseplants tucked into the dark corner of your living room. Birds and rodents scuttle across the rainforest floor, darting around the herbs, orchids, and lichens emerging from the nutrient-rich soil. They too are food for bigger predators: tigers and jaguars, armadillos and pigs.

Ten thousand years ago, more than half of the land on Earth was shaded by tree canopies. Trees offer multiple forms of cooling. Their crowns intercept sunlight and slow the warming of the surfaces underneath. Those cooler surfaces in turn transfer less heat to the air as it blows by. Trees also take up water from the soil and use the energy in sunlight to sweat it out through their leaves, evaporating and cooling the air. The combined cooling effect of shade and evapotranspiration depends on density and scale, but studies from Indonesia suggest that average ambient temperatures in tropical forests are 5 degrees cooler, and summer highs are 15 degrees lower, than those in open areas. That same water vapor can cool and form white clouds in the atmosphere, reflecting sunlight that would otherwise strike the global surface and warm the planet. And of course, trees pull carbon dioxide from the air and sequester it in their trunks. As more of this gas is removed from the atmosphere, more terrestrial heat can escape and radiate out to the cold darkness of space. Although trees' global cooling effects are their most consequential, it is primarily their microclimatic effects that living beings perceive every day.

At the bottom of the rainforest, understory plants thrive in the absence of sunlight. In the near-permanent shade of the tree canopies, the soils stay wetter longer. Growth is slow but consistent. The floras have learned to make the most of what little energy they have. A trademark feature of understory plants is their giant, floppy leaves, which have grown larger to capture more sunlight. The overstory allows no more than 5 percent of sunlight that strikes the treetops to break through and reach the forest floor. Additionally, the leaves are darker and more colorful because they have evolved to absorb a greater range of the energy in that light. The abundance of plant life in this complex ecosystem is made possible by shade.

The smoldering desert could not be more different from the moist rainforest. And yet there's a lesson to take from the forest. Amid the khaki dunes of Arizona, biogeographer Greg Barron-Gafford has constructed an artificial overstory out of photovoltaic panels, emulating the rainforest's shady canopy in the treeless Sonoran landscape. In this man-made shade grow rows of vegetables that would otherwise be burned by the sun of the high desert. Basil, tomatoes, and carrots flourish alongside more traditional crops like beans and chiles. Just a few hours of direct sunlight at sunrise and sunset on top of the diffuse light scattered by the soils around the solar panels at midday are enough to kick-start photosynthesis. Barron-Gafford has a name for this ingenious system of simultaneously growing crops and making clean energy: agrivoltaics.

Crops under canopies? It's a cool idea. In arid regions, gardeners have long used shade nets and canopies to raise their plants. African and Middle Eastern farmers shelter their seedlings under wooden slats, reed mats, and plastic fabrics to preserve soil moisture and prevent sunscald. The Indigenous farmers of Arizona once grew beans and herbs in the shade of wild mesquite trees. Plants need sunlight, but not nearly the amount they're exposed to in bright Arizona. Like animals, they have their limits. When a leaf's surface is too hot, its microscopic pores, known as stomata, close to conserve water. And when they close, the plant also stops taking in carbon dioxide and photosynthesizing. But under Barron-Gafford's synthetic canopies, this does not happen. The crop surfaces stay cool and the plants continue to grow. Herbs in the shade photosynthesize at a slower rate but do not stop, unlike their counterparts in full sun. Like leaves in the rainforest understory, the leaves of the basil plant grow larger, in pursuit of light. The chile peppers produce three times as many fruits, and the cherry tomatoes sprout twice as many fruits as unshaded varieties. Eventually, it leads to higher crop yields. This could be an important discovery on a drier planet.

Moving northeast from the Arizona desert, you will arrive in the vast Midwestern plain. Here, farmers grow not only sensitive herbs and vegetables but, above all, amber waves of grain. These endless acres of wheat and soybeans may also benefit from shade, as do the animals destined to consume them. Cows, pigs, chickens, and other farm animals are endotherms: creatures that thermoregulate internally. Endotherms achieve their ideal body temperatures metabolically, breaking down food and generating the energy they need to live. Heat is a byproduct of this process, and some animals produce too much. The bovines, for example, rely on massive organs called rumens, and even in mild weather, these animals can hardly bear the temperatures of these internal fermentation vats. They get mad and start fighting and mounting one another—bulling. They make less milk. And in extreme cases, they keel over and die. Brad Heins, a University of Minnesota animal scientist, believes more shade could help heated cows chill out. Colorado State University professor Temple Grandin has toured feedyards in Arizona, California, and Australia, and in her travels couldn't help but notice how much happier the cows were under black shade cloth and corrugated iron canopies, calmly chewing their cud in the shade. As Colorado's climate approaches outback-like extremes, she believes such shade will be a necessity.

Smaller endotherms have to find their own shade. Like the ectothermic lizards that they live with, desert rodents avoid overheating by burrowing in the dirt and hiding in rock crevices or taking midday rests under a bush. The burrows are much cooler than the scant shade offered by the northern face of a tree, where their avian counterparts nest to protect their chicks from the sun. Morphological adaptation has proven key to rodents' survival. Rabbits have large, thin ears filled with blood vessels that dissipate their body heat. Some birds protect themselves from the sun with thick, robe-like plumage that blocks the heat before it reaches their skin. And some squirrels have grown longer tails, the better to throw over their backs like parasols and shade themselves. In the summer, they press their bellies on the shady ground to sploot, cooling their flesh.

Cats and dogs, camels and ostriches, gigantic elephants, and our genetic cousins the bonobos and chimpanzees, with whom we share 99 percent of our DNA—all these animals seek shade. Some great apes eke out a tough life on the Senegalese savanna. The dominant apes fight viciously to secure rights to a cool cave or stream, while the adolescents are made to wait on the dark side of a trunk. Compared to forest-dwelling apes, the ones who must scrap for solar shelter have elevated levels of creatinine, a compound that indicates dehydration, and cortisol, the stress response hormone. From a chemical perspective, the primates that live in the canopies are happier, healthier, and calmer.

And unintuitively, it is sometimes easier for apes and other animals to see in the dark of the rainforest than in the blinding sun of the open plain. All vertebrates have a sophisticated visual system with an astonishing range. They can see well on the sunniest day and in the dimmest moonlight, even though one is a million times brighter than the other. The sunrays that reach Earth's surface contain three main forms of energy. Besides visible light waves, which our eyes process in color, there are invisible infrared waves, which we feel as heat. And then there are the ultraviolet waves, which are the most powerful of the three, capable of melting skin tissues and turning them red, wrinkled, or cancerous. Our eyes are especially sensitive to UV, and a direct look at the sun can be dangerous or even blinding, causing corneal burns and cataracts. To protect those tissues, a vertebrate's irises instinctively tighten around the pupils, which shrink to tiny apertures. In a few minutes, the eyes will adjust to the light, but for a vulnerable animal, those minutes may be the difference between life and death. The eye muscles fight to defend against permanent damage, hardening into an exhausting clench. We all know the tiring sensation of squinting in the sun. Shade is literally easy on the eyes.

One More Animal in the Sun

Six hundred miles from the Senegalese savanna lies the bustling coastal metropolis of Dakar. Here you will observe another animal species under the sun: a female human setting out on foot to the market. As she steps out of her apartment onto a busy road, the daylight washes over her and she recoils. She reaches into her purse for a pair of sunglasses to don.

Like other creatures, humans have their methods of beating the heat. Some occur reflexively. When the sun hits our face, invisible infrared waves travel at the speed of light through the skin's surface. Inside, subatomic electromagnetic particles called photons slosh our skin tissues' water molecules around, and that movement warms them up. Depending on the intensity of sunlight, it may only take a fraction of a second for us to perceive this molecular movement as heat. Embedded in these skin tissues are thousands of bundles of free nerve endings called transient receptor potential channels, or thermoreceptors. These nerve endings do not register a specific temperature so much as a changing one. Almost instantly, these molecular thermometers turn on and call the brain with an urgent message.

When we are cold, solar heating can be a pleasant sensation. But by mid-afternoon, when it is added to high air temperatures, the message is received a little differently. The distress signal travels through the spinal cord and up the brain stem to the hypothalamus, the part of the mammalian brain that regulates body temperature. Upon receiving this information, our natural thermostat orders our body to cool down. It tells the blood vessels under the skin to dilate, bringing hot blood to circulate near the surface, where it can transfer heat to the environment. If that doesn't work, then the hypothalamus commands millions of sweat glands to turn on and release a torrent of salty water. The blood under the surface evaporates the sweat and cools the skin. Physiologists say these reflexes are evidence of autonomic thermoregulation: a sophisticated ability to automatically control our core body temperature and dissipate any heat that threatens the preferred internal status of 98 degrees.

But according to Zachary Schlader, a human physiologist at Indiana University, these automatic cooling responses come at a physical price. Vasodilation, for example, forces the heart to beat faster and more vigorously. Blood that normally nourishes the brain and the internal organs can pool in the ankles and feet. Sweating deprives the body of water and makes us tired. Moreover, its effectiveness can be limited in humid weather. Fortunately, humans have other ways to cool down. The woman in Dakar can flap her shirt to fan warm air off her skin. She can step back inside to refresh herself in air-conditioning. Or she can cross that busy road and walk on the shady side of the street. Just like every other animal, humans can behaviorally thermoregulate.

Although autonomic and behavioral thermoregulation are initiated simultaneously, Schlader observes that one usually precedes the other. He hypothesizes that thermal behavior has a protective effect, as preemptive action that humans take to avoid the autonomic responses. In high temperatures, it can take a lot of work to maintain a biologically normal internal temperature. Seeking shade spares us some of the effort. At the same time the thermoreceptors contact the hypothalamus to trigger biological cooling, they fire parallel messages down different neural pathways to the cerebral cortex, the part of the brain that makes decisions. Somewhere along the way, the heat message has split. And while the autonomic message is carried to the hypothalamus to initiate sweating, the behavioral one is destined for the insula. This is the part of the brain that registers the prickling sensation in our skin and turns it into the feeling of discomfort. That feeling is shared with the orbitofrontal and cingulate cortices, the brain's middle management. Do they need to do something about this discomfort? If the answer is yes, the cortices send a new message to the prefrontal cortex, and it's in the Brodmann area 10—one of the most mysterious parts of the brain—that the bosses make a decision. Will they dismiss the recommendation? Or will they send a new message back down the neural pathways?

Schlader suggests that in the long history of animal evolution, the autonomic responses of sweating and vasodilation appeared somewhat recently, largely with the emergence of mammals. Behavioral regulation, however, has been there all along. Some evolutionary biologists have begun to challenge a long-held view that our species emerged on the savannas. Rather, they believe early humans came out of the trees, and for that reason we maintain an affinity for forested areas that offer both prospect and refuge. Ecologists have a name for the feeling of protection that we are afforded by partial shelter: the "forest edge" effect. We may believe that as humans we have an incredibly sophisticated afferent system, but in fact we make decisions based on the same thermal-neural hardware as monkeys, birds, rats, fish, and countless small animals whose core temperatures are tethered to the environment. When the woman in Dakar crosses the street to get out of the sun, she is no different from the Chinook salmon sheltering in a riverbank on the other side of the planet. All animals seek shade to cool down. Perhaps what makes us human is our ability to appreciate the delight.

Researchers have a name for the pleasure we derive from shade. The Canadian neuroscientist Michel Cabanac coined the term "alliesthesia" in 1971, from the Greek words for "changed sensation," after an experiment in which he instructed test subjects to sit in water baths and report their perceptions of comfort. Cabanac also monitored their skin and core temperatures. There did not appear to be a specific water temperature or even a range that was universally perceived to be comfortable. Rather, the preferred water temperature ran lower as the subjects' temperatures rose higher and vice versa. Cabanac concluded comfort was derived from contrast. Coolness never feels better than when we are warm.

Scientists do not know the exact number of thermoreceptors embedded in our skin, but they believe we have many more that are activated by cold stimuli than hot: six or seven times as many cold thermoreceptors as warm ones in our face, twenty times as many on our hands and forearms, and thirty times as many in our chest. Beneath the surface, our flesh is studded with these cold spots: our foreheads, abdomens, shoulders and backs, and thighs and calves. The presence of so many cold thermoreceptors is sometimes used to explain why biting winter winds are more unbearable to humans than a hot desert sun. It may also explain why a woman on a walk to the market can tolerate the humid heat so long as she is in shade.

Shade soothes the senses. When the sun's light is removed from bare skin, the infrared waves diminish and the agitated water molecules in our tissues begin to calm. They slow down, which cools them. As the urgent message of rising heat quiets, different thermoreceptors send a new message to the brain. This message travels the same neural pathways through the spinal cord and brain stem before finally arriving in the insula, where it is coded as comfort, or in sweltering conditions, as relief. Human physiologists marvel at the speed at which this change occurs. Shade instantly lifts our mood, even when we are still too hot, from a clinical perspective. The thermal comfort specialists who study human satisfaction with the environment call the euphoric sensation an "overshoot." In these moments, the psychological impact of shade may be far greater than its actual physiological benefit. The first step on a shady path may not be any cooler than the fiftieth, but it sure does feel that way. In fact, by the hundredth step, we may not even notice the shade at all.

Because humans have a powerful autonomic thermoregulatory system and an enhanced capacity to sweat, it has been suggested that the effects of shade and other forms of behavioral thermoregulation are only skin-deep. Or that they're all in our heads. But Cabanac believed there is a physiological purpose for thermal sensation. Discomfort is a thermal alert. When sunlight warms our skin, it is likely to warm our insides next. And by the same measure, comfort is a thermal guide. When shade cools our skin, it could cool our insides, too. In due time, other physiologists studying shade's effect on humans would bear him out. In the shade, overheated bodies return to equilibrium. Blood circulation improves. We think more clearly. We see better. In a physiological sense, we are ourselves again. The pleasure of shade is the sensory reward for our good thermoregulatory behavior, and an indicator that we are taking care of ourselves. Humans are just like salmon, soybeans, and salamanders when they cannot access shade: Things start to go wrong.

Chapter 2

Shady Lanes: The World's Oldest Cities Know the Wisdom of Shade

Four thousand years ago, a wandering Sumerian dragging a donkey across a marshy plain looked to the horizon. His pulse began to quicken. In the distance were the hazy contours of a mound rising from the earth: the ziggurat of Nanna, a monumental temple that shone in the sun. The massive fortified walls, spanning two hundred feet across, were scaled by long, buttressed stairwells that culminated in flat sky terraces. Perhaps the grandeur inspired awe or made the traveler tremble in fear. A mind-melting sun, bobbing overhead, could curdle a man's thoughts and make him question his sanity. In the Mesopotamian flats, temperatures soar into triple digits, and in the cloudless summers, when the sun scorches the grasses and dries ponds into spittle, they can easily clear 130 degrees.

More likely, the sight of the ziggurat triggered immense relief, because it meant the traveler had finally reached a city. Across the Mesopotamian plain, there were scores of urban settlements where people hid from the sun behind thirty-foot gates. Soon enough, the traveler hoped, the moon god's followers would swing those gates open and welcome him with a whoosh of damp and cool air gusting from a twisting warren of muddy alleys and footpaths. It wouldn't be long, he figured, until he could enter the city of Ur and refresh himself in its shade. "It's one of the reasons that cities really took off, I suspect, in this region quite early, is that they just were such a nicer living environment than dispersed small villages," said Mary Shepperson, an archaeologist who has spent two decades excavating the ancient cities of the Middle East. She even lived among the remains of the great city of Ur, one of the first of human civilization. "There's so much shade and so little direct sunlight."

Forget palm trees and ponds. In ancient Mesopotamia, cities were the real oases, the places where weary travelers got out of the sun. Long before anyone dreamed of a street tree, the Sumerians made shade from the city itself. They did this mainly by packing their houses close together. They built squat, ten-foot-tall homes that shared two or three walls with the neighbors. A snaking maze of alleys no more than five feet wide provided ingress. Only when the sun was directly overhead did the light break through. Mesopotamian streets were deep, narrow canyons, at least twice as tall as they were wide. They were plunged in shade.

Undoubtedly, the density of Ur was a function of material necessity. A smaller, more compact area was easier to defend. There weren't many trees on the plain to use as fuel to make mudbricks or use as lumber. But the Mesopotamians surely grasped the climatic benefit of shade. The close quarters were far more pleasant than the arid countryside outside the gates, where a destructive summer sun crumbled bricks and ended human and animal life alike. "Nothing lasts very long at the surface in southern Iraq," Shepperson explained. "Every year, the rains come, the salt crystals dissolve in the water, and they get into the bricks and seep into the pottery. Then, when the sun gets on them, the water evaporates, and the salt crystals grow and basically explode everything." On the plain, the sun would have blasted the bricks to bits.

Archaeologists have long believed Mesopotamian cities were oriented on diagonal grids to catch prevailing winds and ventilate the streets. Shepperson has another theory. When she was an illustrator on a summer dig in southeast Turkey, she and her colleagues dragged their desks outside to search for fresh air, only to become nomads in a daily migration for a different sort of relief. The sun was miserable and the August heat was intolerable. They ended up circling the compound by moving their chairs around to follow the shadows. It was easy enough to do. But as she discovered on the dig, outdoor seats were once benches built into the ground. It opened her eyes to the importance of shade.

Today we take urban grids oriented to the cardinal directions for granted, but we hardly appreciate their climatic effects. The buildings on a street running perfectly north–south throw their shadows to the east or the west. While façades on one side are cool in shade, those on the other side are warm in sun. On a perfect east–west street, there's practically no shade in the middle of summer except for the consistent sliver thrown by buildings on the south side. A diagonal orientation, on the other hand, offers equal amounts of sun and shade on both sides of the street throughout the day, for buildings and street surfaces alike. Shepperson used a solar data calculator to model the sun's daily and seasonal paths over Mesopotamian cities. Between the street orientation, the narrowness of the thoroughfares, and small protrusions like vertical rooftop parapets and horizontal eaves, Shepperson believes Ur and other ancient cities offered pedestrians a pleasant network of shadows to traverse the city out of the sun's glare.

The most effective shadow networks offer more than personal protection from the sun. They also cool the air. Shepperson emphasizes two of the main factors that create effective shading in a city: solar orientation and geometry. When streets are oriented along a diagonal grid, all surfaces are shaded at some point during the day. And when the buildings are taller than the street is wide, the surfaces remain in shade longer. Together, these two factors can result in minimal solar heating of urban streets. But keeping the air itself cool requires confining it in the shadows. In the medina of Fez, Morocco, some buildings are ten times taller than the narrow streets are wide, creating an effective enclosure. The bottoms of the street canyons remain in shade all day and do not warm to heat the air above them. Dense construction then barricades this cool air inside. As a result, the air in the medina can be almost 20 degrees cooler than the air beyond the medieval walls. In mild climates, we might describe such enclosed environments as claustrophobic. But in cloudless lands, the shade of density can be a welcome reprieve. It may be why compact urban form endured in the deserts of the Middle East and northern Africa for almost five thousand years.

Shade was once so closely associated with urbanity that in Sumerian mythology, the gods describe its creation as one of the essential purposes of cities: "Let them build many cities so that I can refresh myself in their shade." For the ancients, shade was more than a physical phenomenon. Practical, climate-responsive design had cultic implications. A house that was shady on the outside was cool and temperate, but also lucky. A house in the sun was hot and uncomfortable, but also cursed. Only the temples, where gods reigned, should be sunlit. "There's not a single window that's ever been found in an ancient Mesopotamian house," Shepperson told me. The only source of light was the doorway to the courtyard, where cooking, working, and living were done. The house was for sleep and the occasional meal. There was no need to invite the sun inside.

The idea of basking in sunlight, even during a cold, wet winter, was taboo to the Mesopotamians. It was a somber ritual performed in the temple courtyards and on the glaring terraces. Shepperson suggests that by the second millennium B.C.E., the temples may have been constructed to face the rising morning light so the sun god Shamash could better judge his dominion. Mesopotamians worshipped the sun, but not in the way we do today when we strip down for a tan. Shepperson believes the ordinary Sumerians revered the awesome power of sunlight and therefore preferred to avoid it. Unfortunately, that was not possible on the broad imperial roads that Assyrian and Babylonian rulers began to cut through their cities in the first millennium B.C.E. as demonstrations of godlike power. The great swaths of light were said to "shine like the day," and the lack of shade would have frightened anyone who attempted to cross without sun protection. Of course, the kings suffered no such discomfort, because they were trailed by servants shading them with parasols.

The wisdom of shade persisted in hot regions for millennia. It was known to the Syrians and Phoenicians and Persians, embraced by the Greeks and Romans, and eventually exported by the Islamic caliphate to the land that became modern Spain and Portugal. Yet it was never embraced by the Europeans farther north, who had their own theories about healthy cities and how to build them. They carried these theories with them as they expanded their empires to new lands. In the temperate climates, the sun was their friend, not their enemy. They had little use for shade.

Public Shade by Official Decree

There is no place like Bologna, the Italian city that boasts thirty-eight miles of covered sidewalks, many of them carved right out of the ground floors of buildings. The residents call these public passages *portici,* and they have sheltered pedestrians from summer sun, driving rains, and winter snows for a thousand years. The longest portico of all, a freestanding holy path that leads from the center of town to a hilltop church, offers two and a half miles of continuous shelter. It is possible to spend an entire day tunneling through Bologna's porticoes, and the effect is enchanting: streets of endless arched columns and vaulted roofs, rhythmic shafts of light and shadow that seem to go on forever. "It's an improbable city, Bologna," the critic John Berger wrote. "Like one you might walk through after you have died." Unlike the narrow and enclosed streets of the ancient cities, the porticoes do not have a detectable effect on air temperature. They are too open and the air inside mixes rapidly with the hot air outside. But even this mixing has a thermal effect, in the form of a vacuum that is created where the high-pressure air of the sunny street meets the low-pressure air of the shady footpath. We call this a breeze.

Architectural historians say the portico originated in ancient Greece as a sanctuary called a *stoa*. In its simplest form the stoa consists of a long back wall and a simple roof supported on a row of posts. Later the stoas became monumental stone colonnades that surrounded the town square in the agora, the beating heart of the Athenian republic. It made sense for a government that aspired to transparency to conduct its affairs in a transparent building. These open halls became the places where civic activities occurred in full view of the people—the convening of the popular assembly, the hearings of the magistrates, the displays of laws and art, and the transactions of artisans, merchants, and businessmen.

Because the stoa offered shelter from the elements, it was also a natural venue for gathering. It was the urban meeting place where people struck up conversations and took in Hellenistic education. The great philosophers made their homes in the stoa—literally, Diogenes slept there. Socrates strolled in its shade. So did Zeno, whose Stoicism school was named after the Stoa Poikile (Painted Porch) where he lectured and found his audience. In the fourth century B.C.E., when Aristotle sought a venue for his philosophical discoursing, he began lecturing in a public gymnasium called the Lyceum, and under the cover of its *peripatoi* (shady walks) founded the school of the Peripatetics.

From the Greeks, historians say, the Romans were inspired to colonnade the façades of the buildings in their grand forums, and the roads that led there, too. Rome's streets were muddy, messy, and washed out. The city of walkers likely appreciated the decision to separate the footpaths, rebuild them with paving stones, and put them under roofs. A typical covered sidewalk had a row of stone or marble columns topped by a flat or triangular roof, maybe ten to thirty feet high, that was usually made of wooden beams and covered by terracotta tiles. The whole structure clung to the surface of the flanking building.

The first to sponsor these street improvements were the public officials of the Roman Republic, but it wasn't long until the emperors took to building one portico after another in a bid for the public's affection. Augustus, said to have found Rome a city of bricks and left it a city of marble, claimed credit for about a hundred new porticoes. Nero was forced to add more after the destruction of the great fire of 64 C.E. Initially, he proposed fighting future fires with sprawl, preventing their spread by constructing wide, gridded roads and leaving space between buildings. "And people hated it," explained Diane Favro, a University of California, Los Angeles, scholar of Roman urbanism. Unlike the ancient city's narrow, winding alleys, Nero's broad new thoroughfares would offer no shade. The emperor reluctantly returned to the colonnade as the pragmatic expedient.

In ancient Rome, the porticoes formed a shadow network that stretched for two miles, and a fully covered walk to the Roman Forum struck visitors with wonder. One could effectively stroll from one side of the ancient city to the other in shade. The porticoes had an obvious appeal to the emperors as a fast, cheap way to impose architectural order upon a chaotic city. But the beauty of so many perfectly spaced columns was probably the last thing on the minds of ordinary Romans. Not least because their apartments were so dismal, their lives happened outdoors, and they went about their business in the shade of the porticoes, chatting with friends or waiting at the *taberna* for their meal. There was no rush. The porticoes were a fine place to pass the time, and as evidence, archaeologists point to game boards they have discovered, scratched into the footpaths and steps. People-watching was another popular amusement. "You know, there's not that much entertainment," Favro reminded me. Sitting in the shade was one of the few pastimes available to them.

Not all shade was created equal. There were the magisterial porticoes of the official order, and then there were the plastered brick arcades of the shops, erected by merchants to shelter areas outside their bars, restaurants, and stores. In a pinch, shopkeepers rammed a wood pole into the ground and strung a cloth awning to a nearby window. Any way they could make shade, Favro said, they did. "You did it until somebody said you couldn't." The awning created more protected space to show off the merchandise, but the shade was the attraction, drawing in customers who might have taken their business somewhere else. The benefit was obvious. "I would rather go to the stall where I could stand in the shade in line to get my bread, than the one where I had to stand in the sun," Favro said. In a warm Mediterranean climate, "it's just kind of a no-brainer."

Sometimes, the space between the columns was rented to a merchant who didn't have a permanent shop, like a street vendor today. They scratched out a living by selling jewels, sandals, and poems on the sidewalk to throngs passing by. The more desperate members of Roman society sold sex in the intercolumniations, and some may have even slept there. Rome's media elite despised the urban chaos, and it is no coincidence that the playwright Plautus had a special slur for the drunks and gamblers who populated his tales: the *umbratici,* literally "shady people." The poets preferred to stroll the mosaic sidewalks of the magnificent porticoes that emperors and generals built in the forums, temples, and basilicas in the center of town. In these Hellenistic pleasure walks, where bronze busts and spoils of war were proudly displayed between the columns, Catullus, Ovid, and other poets met shady ladies of their own.

Rome was the capital of an empire, and as it expanded into the Middle East, so did its architecture. The generals cut broad thoroughfares into conquered cities and framed these roads with grand colonnades as emblems of the Pax Romana. As in the capital, the structures served practical purposes for shopkeepers and customers, and wealthy patrons who used the porticoes to advertise their largesse. But more than that, the porticoes were a true public space, a neutral arena that flattened differences of class and privilege. They were the places where everyone in town could be found. And because they were covered, the mingling could happen all year-round, rain or shine. Public "association is not altered by any season," bragged the orator Libanius in 360 C.E., three centuries after the Romans built a grand colonnaded street in Antioch. In other cities, inclement weather banished residents to their homes "as though they were prisoners." But in his city, "friendship grows by the unceasing nature of our association."

If the great benefit of urban living was all the "meetings and mixings with other people," as he believed, then the quintessential urban architecture was the covered sidewalk.

It may be hard to imagine, but the ruins of columns still standing in some ancient Roman cities once supported canopies of rafters and tiles. To experience the scale of these original urban forests, one must visit Bologna, which rediscovered the portico in the medieval era. In the thirteenth century, the city convulsed under the stress of a modern burden—a housing shortage. The students who flocked to the university from all over Europe had nowhere to stay. Bologna used to be a city of crenellated towers, but for public safety reasons the authorities capped the heights of new buildings. Unable to build up, the Bolognese instead built out and extended the second floors of their townhouses over the street. Under papal rule, most Italian cities abolished the commandeering public space for private good. But perhaps in recognition of the benefits of shade, the Bolognese came to appreciate these odd projections. In 1288, the city not only legalized these ramshackle porticoes but made them compulsory by statute. From then on, every landowner in Bologna's busiest districts was ordered to furnish a covered public passage and maintain it for eternity at their own expense.

Bologna's porticoes became outdoor workshops where artisans, bakers, and carpenters worked in natural daylight without bearing the brunt of the sun. Because Bologna was a university town, the locals surely appreciated the porticoes as the Greeks did their stoas—as open-air salons where conversation and debate sparked in public. Dante, Petrarch, and Copernicus are a few of the students who strolled in their shade. So did the Renaissance architect Leon Battista Alberti, who adored the porticoes and advocated for them around piazzas. To this day, the corridors are jammed with students hanging murals and banners in the rafters. The porticoes encourage civic virtue. In most cities, it would be difficult to force a developer to meaningfully shade the sidewalk without getting anything in return. But in Bologna, it has been the order for centuries. The porticoes are what one critic called the physical expression of the residents' social solidarity, "altruism turned architecture."

Near the end of the third century B.C.E., after having conquered the Mediterranean, the Romans took control of a feverishly hot port town near the southern coast of the Iberian Peninsula. They named the town Hispalis. Seville, as it's now known, is a place where keeping cool is a major preoccupation. The people here have a dance for hand fanning: flamenco. They bid friends farewell by wishing them a path in the shade: *Vete por la sombra!* And at the bullring, seats in the *tendido de sombra* are a hot commodity. The tickets cost two or three times more than those in the sun.

"Seville is a city of shadows," the English writer V. S. Pritchett observed. There are the dense shadows in alleyways of the oldest districts, winding under trellises of bougainvillea. There are the drowsy shadows of the courtyards, fragrant oases that are festooned with azaleas and orange trees and cooled by splashes of water evaporating off ceramic tiles. There are the cave-like shadows in homes, which are boarded up to create the cool darkness needed for an afternoon siesta. A traditional Spanish home has thick walls that slow the intrusion of daytime heat and preserve the cool overnight air. Small windows are punched through to let in the minimum necessary light. In Seville, practically all those windows are shaded, if not by wood or bamboo roller blinds, then by the tranquil shadows of the *persianas de esparto,* threaded grass curtains that are draped gently over wrought iron balconies.

And there are the dreamy shadows of the *toldos,* curtains that billow like white linen bedsheets in the city center. Every spring, crews arrive with cranes to hoist them to the upper stories of buildings, stringing them over the streets to shelter the pedestrians from the impending arrival of a high summer sun. They drape over the busy shopping district and the quiet residential side streets. They sail over bus stops and taxi circles. In historic plazas, they are fastened to metal poles and spread like a canopy on the frame of a four-poster bed. The toldos have cast benevolent shade for the public for more than five hundred years, and Sevillanos are rightfully proud of them as the traditional architecture of the public realm.

Fabric canopies (toldos) *have long shaded the streets of Seville.*

Like the porticoes of the Roman world, the toldos do not meaningfully cool the air of Seville. The textiles are too thin to completely occlude the sun. The small fraction of cooler air in their shade is swiftly displaced by the dry and hot air that gusts through the streets. The toldos' thermal effects are more subtle. They dull the glare on the street, easing the strolls. They throw feathery shadows on buildings, cooling the surfaces. And they soften the sun's prickle on bare human skin, preserving the languid comfort of an Andalusian summer.

No one really knows where these canopies come from. Given Seville's Roman heritage, the answer may seem obvious. Although the Romans were known for their porticoes and colonnades, they also appreciated awnings. Julius Caesar arranged for a massive and luxurious silk to cover the Via Sacra to shade Romans watching gladiator fights. According to one ancient source, the spectators were more amazed by the awning than the combat. His successor, Augustus, did him one better by covering the entire Forum in red, yellow, and purple canopies, flapping and fluttering over the theaters. The best-known awning is the velarium, a system of retractable fabric sails rigged to giant poles in the Colosseum. The sails were said to make a cracking roar when the wind beat them about and smacked them against the beams. This primitive technology must have been a spectacular sight to plebeians. Besides the many frescoes and writings depicting the awnings at the Colosseum, inscriptions in Pompeii that announced games at the town's amphitheater advertise the shade as its own attraction: *Cruciarii, venatio, et vela erunt* ("There will be crucified people, a hunting spectacle, and awnings").

But Seville's awnings are equally likely to have been introduced by the Moors. In the eighth century, Muslim armies invaded the peninsula from northern Africa, and they set up dynasties that ruled the region for the next seven hundred years, renaming it al-Andalus, or Andalusia in Spanish. In this arid climate, they reapplied the architecture of the Arab world. Their homes were designed to be naturally cool, and located along narrow, shady lanes and alleys. Their outdoor markets, or souks, were roofed, covered by either towering vaults or simpler and shabbier canopies of mats, reeds, and fabrics. Over time, this origin story goes, the street coverings were replaced by canvas sails that were discarded from ships docked on the river, like those that ferried conquistadores to the New World.

Still others insist it was neither the Romans nor the Moors who introduced the Andalusians to the pleasures of canopied streets, but the Castilians who conquered the city in the thirteenth century. However the practice arose, shade has been a part of Spanish life for centuries. According to writing from 1454, before the Corpus Christi festival in late May or early June, Catholics strung billowing awnings and tapestries over the streets to shelter the priests' procession. City records show that toldos were arranged for funerals and other religious rites as far back as 1570. And photographs from the nineteenth century show shopkeepers on Seville's market streets suspending the canopies from their rooftops for the benefit of their customers. The shades stayed up until evening, when they were retracted to allow cool air to flow, and for that reason they are sometimes called *velas* (candles), because they go out at night. It is not clear when the authorities took them over, but at some point, the toldos became a municipal service, and their shade a civic resource that Seville and other Andalusian cities provide for free.

Today, Seville's government shades not only the streets in the city center, but also two dozen schoolyards. This may be because Europe is the planet's fastest-warming continent, its temperatures rising at twice the global rate. Spain is one of its warmest countries, and in the summer, Seville is frequently its hottest spot, with months of soaring air temperatures. Chronic heat is dangerous, because it denies our bodies the opportunities needed to cool down and recuperate. Any relief that can be afforded, the better.

I visited Seville in July 2022, just in time for the arrival of a stretch of heat so punishing the city government partnered with scientists to give it a name, Heat Wave Zoe, in a quixotic attempt to communicate the seriousness of the situation. Daniel González Rojas, a scruffy, card-carrying Communist city councillor, dismissed the effort. He was more concerned about shade. We left the city center to head to the Triana Bridge, one of nine in inner Seville that span the Guadalquivir River and its canals. Like the others, the unshaded crossing was difficult to bear. As the pavement burned through my soles,

González Rojas explained the unfortunate choice that Seville's pedestrians faced: either a painful and intimidating passage on foot, or a comfortable drive that made the heat even worse, both in the short term through belching tailpipes and in the long term with carbon pollution.

González Rojas whipped out a black hand fan to cool down. As we came to the middle of the bridge, squinting in the sun as it glinted on the metal railings, he pointed out garland lights hooked to lampposts for that evening's summer fair. If it was so easy to hang these up, then why not some toldos, too? He explained the barrier was administrative. Historic bridges like the Triana required special considerations for new designs, and the city dragged its feet on approvals. It now seemed Sevillanos could no longer wait. Climate activists had resorted to bringing their own shade to a nearby bridge, hoisting miniature toldos and green umbrellas over elderly pedestrians to call attention to the need. Cultural patrimony is important in Seville; it's why the government still erects toldos in the center. But if leaders want the rest of the city to be usable in the future, they may have to approve some changes to the bridges. The toldos that González Rojas and others have demanded are a possible solution.

The toldos have been upgraded over time. Darker, thicker cotton that soaks up sun has given way to brighter and thinner synthetic fabric that reflects it away. The municipality covers more streets than ever before, and because summer arrives earlier and ends later, the shades stay up longer. But González Rojas is not the only one who feels these efforts are insufficient. He is joined by another city councillor named Álvaro Pimentel Siles, a conservative lawyer from a wealthy political family. "I studied here, I was born here, and I will die here," the politician informed me, banging his fist on a conference table in his city hall office. An assistant handed me an orange paper fan stamped with a slogan from a campaign that his party had waged the previous summer to put pressure on then-Mayor Juan Espadas. "Seville doesn't deserve more messes," it read in Spanish. "Where are the toldos, Mr. Espadas?"

An unlikely coalition had formed in the shade of the toldos. On one flank were the leftists, like González Rojas, who wanted climate justice for working people. On the other were right-wingers like Pimentel who looked out for business interests. Although they disagreed on most everything else, on this matter they were united: Seville needed more shade. "Seville is a city where people spill out onto the street," Pimentel told me through a translator, mentioning the fairs and religious rites that happen outdoors and beautiful boulevards where people *pasear*. He didn't believe the torrid summer temperatures were unusual. It's always hot in Seville, Pimentel told me, brushing off the implication of climate change. Nevertheless, when the heat lasts as long as it does, "the temptation is for people to go to the big commercial centers, which are fully air-conditioned." Pimentel would rather keep them comfortable in the streets.

In 2021, Seville residents were disturbed when there were no toldos at all. At first it seemed the installation was merely delayed. But as May rolled into June and temperatures rose, the dread mounted. City officials blamed a bureaucratic snafu with a vendor. By July, the local shopkeepers were fed up. Their ally Pimentel took to the newspapers, blasting the mayor for the scandalous *verano sin sombra*. The world was emerging from a dreadful pandemic and the lack of shade in Seville hindered its economic recovery. The negligence was so staggering that Pimentel's political rivals called for the mayor to resign. How could the hottest city in Spain not have toldos? It was a "total shame," Pimentel told me. He fears that in the future, if Seville is too hot to walk, then no one will visit the shops. And if that day comes, the desolate streets will not only impact the municipal coffers. They will mean that Sevillanos have been forced into doing the least Sevillano thing of all: staying inside. "We're just trying to make sure that the traditional businesses still have a chance," he reiterated. "We can't air-condition the streets. But we're doing what we can." In the face of crushing heat, Pimentel is fighting to preserve Seville's outdoor way of life.

Shading the New World

Long ago, the Spanish brought this way of life to the New World. In the sixteenth century, King Philip II of Spain issued a set of ordinances for the expanding empire. The Laws of the Indies, as the 148 ordinances are known, are dominated by economic and political decrees, yet a concern with adapting Spaniards to new climates was woven throughout. The monarch ordered that all new cities be centered on a public plaza, and that this plaza be surrounded by *portales,* their version of the Italian porticoes. These porches would be designated for the "considerable convenience" of the merchants who assembled there to sell their wares. From the plaza would proceed the colony's main streets, and those too would be porched.

The ordinances also decreed that in warm climates, the gridded streets of new cities were required to be narrow and shady. This explains the cozy paseos and corridors that survive in the old quarters of St. Augustine, Santa Fe, and Havana, among other Latin American cities. The Spanish planners pointed the corners of the plazas to the cardinal directions, and oriented the streets diagonally. Just as in ancient Sumer, the city layout secured a constant balance of sun and shade most of the day. This is still the case in Lima and downtown Los Angeles, whose Spanish heart is canted against the rest of the city, which conforms to early America's right-angle grid.

Even as the Anglos took over, the Spanish influence lingered on the American frontier. In time, the *portales* evolved into the dusty wood boardwalks that fronted saloons and hotels in the Old West. Like the stoas and porticoes of yore, these shady porches were some of the only social spaces around, where townsfolk traded gossip while they waited for the mail rider. Today, those places would be all gone but for the historic landmark districts of Gold Rush towns and the sets of Hollywood westerns. They also survive in degraded form behind seas of parking lots, in the thin posts and stuccoed canopies fronting the windows of strip malls. We ditched the public porch to settle under a different kind of urban canopy.

The New World was once thick with shade. Think of the great forests of the eastern United States and the jungles of Central America, sheltering plants, animals, villages, and sometimes massive human settlements. Or the stunning cliff dwellings of the ancestral Puebloans, cut into huge niches that blocked the overhead summer sun and admitted the low-angle winter rays. In these physically constricted spaces, the decision to work inside the cave or out, on the left side or the right, was likely determined by the path of a sweeping shadow or single shaft of light.

In the deserts of the Southwest, where extreme temperatures were a constant, the descendants of the Puebloans opted for thermally massive construction. Like the Moors and the Spanish, they built to buffer the daytime highs and overnight lows, and like the Sumerians, they located their homes close together to minimize exposure to the sun. The Pueblos of New Mexico stacked their dwellings vertically, five or six stories high, and entered through the roof. The Hohokam of Arizona, who lived under a burning sun, dug their dwellings into pits, using the earth itself as an insulator. Along the river valleys of the Sonoran Desert, they built a system of canals so effective that modern irrigation networks still follow its path. The Salt and Gila Rivers ran all year, and people jumped in for cooldowns. They grew ferny paloverde and mesquite trees along the banks for lumber but saved bosks for shade. Like the Spanish on siesta, the Hohokam knew better than to work in the middle of the day. If they absolutely had to, they did so under a ramada, a wood-framed structure covered with desert brush and saguaro ribs. Some were light and portable, like the tents we might see today on a construction site or farm field.

Despite shade's evident importance, the New World's most influential colonists, the British, weren't really into it. The weather in the northern colonies wasn't too different from that in England, marked by long, bitter winters and short, warm summers. They were more concerned with staying warm than staying cool. In the southern colonies, where the sweltering heat seemed to last forever, the preferred adaptation of the wealthy was to retreat indoors, lie down in a breezy hallway, and let a slave bear the brunt of the sun. Some wealthy colonists privatized the public Renaissance portico as the entrance to their homes. They had come here to farm and raise crops, not to build arcaded squares.

In the seventeenth century, William Penn came to the New World to build a new kind of city. He was fortunate to have been abroad when the Great Fire of 1666 burned his hometown to the ground. London was a cramped and congested medieval warren where hundreds of thousands of people were packed behind rotting Roman walls. The sanitation system was nothing more than an open gutter of filth that spewed down city streets. Plague was endemic and spread rapidly in overcrowded dwellings. So too did the fire, blazing through a dense mass of wooden buildings. Density did not offer protection from the elements in London, as it had in Ur. Instead, it bred disease and destruction.

In America, Penn and his surveyor unveiled their plan for Philadelphia, a city that would be everything London was not. Instead of a winding maze, it was plotted on a uniform grid. Rather than narrow alleys shaded by their buildings, the roads would be broad, fifty-foot streets and hundred-foot-wide avenues. They aimed to fight fire with empty space, and to combat disease through rurality. Each quadrant of the city was planned around an open square where residents were free to graze their own cattle. Each house was to be placed in the middle of the plot to leave space for personal gardens. This was not a city in the style of London, but a "greene country towne, which will never be burnt and always wholesome," as Penn declared. The plan was much admired and emulated in the broad roads of future American cities like Savannah and later Washington, D.C. There, Thomas Jefferson planned street widths that swelled to 120 feet and buildings that stayed short in pursuit of a city that was "light and airy." Although the ancients came together to shade themselves from the sun, the English colonists who feared fire and disease would understandably do no such thing. Were it not for the emergence of a new kind of urban architecture, we would still be feeling the thermal effects of their urban design today.

Europeans arrived in the New World with a deep and abiding fear of forests. The woods were pagan and unholy places, the realm of savages—a word derived from the Latin for forest, *silva*—and havens of Robin Hood–esque highwaymen. The city was civilization; everything beyond its walls was not. The colonists wasted little time clearing America's vast woodland. Before contact, trees covered the entire Eastern Seaboard. Within two hundred years, about half of them were gone. A few were spared on farms for practical uses; a stand of trees served as a windbreak, while a single mighty oak loomed over a homestead as a lightning rod and shelter from the sun. But in cities, they served no such purpose.

As a result, shade was scarce on the streets of the early northern colonies. When the word was used in a Boston newspaper in 1722, it referred to clouds, not trees. In poems, it was a romantic emblem of rural life. Consider the many good reasons not to plant trees in cities. In arid regions, they require water that may be rare. In medieval cities, there was very little extra space, and already plenty of shade cast by the buildings and porticoes. Their roots wreak havoc on pavements. Branches can break and fall on a pedestrian or a roof. The colonists kept their streets bare for the better part of two centuries because, as an English patrician proclaimed in 1771, "a garden in a street is not less absurd than a street in a garden." A city dweller who planted trees in their front yard was just another country bumpkin.

Architectural and environmental scholars have identified distinct forces in the late eighteenth and early nineteenth centuries that spurred the acceptance and then promotion of city trees. One such force was fire insurance. Conflagrations were once common in the cities, and tree branches and foliage were believed to fuel them. Philadelphia's leading insurance companies refused to issue policies to homeowners who planted wood and even lobbied the state legislature to make trees illegal. But with the rise of brick construction, and the creation of municipal fire departments, the threat of urban infernos was no longer so severe. According to landscape researcher Anne Beamish, a breakthrough moment occurred when the Insurance Company of North America removed tree stipulations from its policies. This freed urban dwellers to shade the streets.

Another force was war. The earliest American town squares were functional commons where livestock was raised, human waste was dumped, and state militias were mustered for battle. Boston Common was totally denuded but for one lone elm. Philadelphia's five public squares were similarly utilitarian and lacked elms and maples. Even the squares of Savannah were free of their now-signature mossy and majestic live oaks. The architectural historian Michael Webb locates the War of 1812 as a turning point. After the militias failed in combat, they were disbanded, and the United States created a standing army. The decision cleared the way for the trampled New England common to become a place for leisure and gradually bloom into a verdurous park.

And a third force was yellow fever. It once swept New York, Boston, Baltimore, and other colonial centers with such regularity that a Charleston physician named David Ramsay called it a "disease of cities." The worst outbreak was in 1793, when sweltering heat suffused Philadelphia with swampy moisture, transforming the city into a breeding ground for the mosquitoes that we now know spread the virus. Within months, 10 percent of the population of the nation's then-capital were dead. Many more had fled for their lives. Some believed that if epidemics could not be controlled, the cities would be deserted and America would be reduced to ruin. At the time, leading physicians like Ramsay abided by an ancient Hippocratic belief that diseases were caused by miasma, or bad air, that arose from stagnant marshes, standing water, and urban refuse, especially when they were heated by the sun. Miasma theory provided a clear rationale for shade as a public health intervention.

But what kind of shade? Noah Webster, then a journalist and scrupulous collector of public health data, was among the first to recognize that New York City, where he lived, was warmer than the nearby rural environs. In the winter, when the air in town was 40 degrees, thick ice formed on ponds a mile away. Over a year, he recorded ambient temperatures that were consistently higher than those in the country. Today, we recognize this meteorological phenomenon as the urban heat island effect. Webster did not believe this was a mere curiosity. He believed the elevated temperatures could weaken the body and render it prey to solar heat. "Another thing to be observed in summer, and especially in time of pestilence, is the guarding the body, but by all means, the head, from the direct rays of the sun," he wrote in 1799. "It often produces sudden death, by means of an apoplexy." One defense against a "stroke of the sun" was an umbrella, "an excellent invention." But that was an individual prescription, not a public cure. To cool the city, he thought that streets could be narrower, and the buildings taller. Webster had rediscovered what Sumerians and Romans had known millennia before.

Charles Caldwell, like Webster a miasmatist, also suggested that urban heat and its associated diseases could be ameliorated with Mediterranean architecture and urban planning. In 1801, the Philadelphia doctor urged his readers to build homes like the Spanish, with thick walls, small windows, and spacious rooms with high ceilings. He suggested cramming tall buildings along twenty-foot-wide roads. Caldwell was the rare dissenter who lamented the celebrated plan for the future capital of Washington, D.C. He predicted the city would be uninhabitable in the summer. "The streets are too wide, and the buildings too low, to furnish any protection against the solar rays," he wrote. "The summer temperature of its atmosphere will be but little below that of the inhospitable desert" of the Sahara.

But because plagues spread in dense cities, many Americans

were conditioned to fear compact urban design. Eighteenth- and nineteenth-century travelers to the Middle East were disgusted by the tight construction. They believed the narrow alleys of Damascus and Constantinople were not sensible responses to local climate but overcrowded cesspools that abounded with foul smells. They scoffed at the awnings in the souks, deriding them as ragged mats that made streets damp and gloomy. Herman Melville cursed Jerusalem's air "pent in" by medieval walls, and another traveler to the Holy City noted that the shady streets appeared to inhibit ventilation. Webster agreed. "This is to embrace one evil, in shunning another," he wrote. Narrow streets blocked the sun, but they also blocked fresh air. He offered a more salubrious design solution.

"Wide streets, bordered with rows of trees, would be infinitely preferable to all the artificial shades that can be invented," Webster wrote. "Trees are the coolers given to us by nature." He was not alone in this recommendation. Doctors came to urge residents to plant trees wherever possible to thwart the spread of yellow fever. Sanitarians recommended them to ward off other urban contagions, like cholera and tuberculosis. These miasmatists ascribed to trees an even higher power than cooling. They believed foliage absorbed miasmas, taking in the bad particles and disinfecting the air. Heeding their advice, European city planners and engineers built wide, straight streets where modern sewers ran below and canopies arched above. In the 1830s and 1840s, the Comte de Rambuteau, prefect of the Seine, oversaw the reconstruction of nearly two hundred miles of roads in Paris and initiated a plan to line new boulevards with drinking fountains and colonnaded trunks. "'The mission with which you have honoured me implies a great obligation, which may be summarised in a few words: *To give the Parisians water, air, and shade,*'" Rambuteau later wrote, recounting what he'd said to the king of France. "Such, in fact, was my programme, my constant thought, the goal of all my labors."

Some land set aside to absorb miasmas became today's tree-shaded parks. Modern New Yorkers flock to the naturalistic Central Park seeking beauty and tranquility, but the park was initially cherished for a different reason, as the "lungs of the city" where New Yorkers could breathe clean air. The park's designer, Frederick Law Olmsted, was right that trees capture harmful gases, although not the ones he was thinking of. Though we now subscribe to the germ theory of disease, we still accept Olmsted's social justifications for his urban woodland. In 1858, when Olmsted was awarded the park's commission, New York's wealthy and leisure classes dealt with summer heat by decamping to countryside homes in the Adirondacks. The working classes did not have this option and suffered in their crowded tenements. Olmsted designed Central Park, and other urban parks in the United States and Canada, to be a cool summer retreats for those who could not afford to leave the city.

The nineteenth century was also the era of westward expansion. As American pioneers fled congested cities and forested countrysides to the Great Plains, some thought the open land was unattractive. No mountains and no trees. One such émigré was Julius Sterling Morton, a Michigander who moved to Nebraska in 1854 to start a family and pursue a career in politics. Morton was troubled by the landscape, which he considered an unproductive plain. He yearned for his native shade. "Almost rainless, only habitable by bringing forest products from other lands," he complained. Morton urged people to improve the land and join a "battle against the treeless prairies." In 1872, Nebraskans commemorated Morton's first Arbor Day celebration by afforesting the state with one million trees. The effort caught on, and Arbor Day became a civic holiday, even in cities.

Arbor Day rose as America's natural forests continued to fall. The clearing of so many trees for farming and lumber was a growing source of national remorse. Perhaps in naturally sylvan regions, reforestation was a necessary salve. But from an ecological perspective, the imposition of trees on the plains was questionable, and in the arid Southwest, irresponsible. The focus on afforestation seemed to preclude other environmental improvements and ways to make shade. It would not be long before frontiersmen were planting the arcaded plazas of Santa Fe and New Orleans, erasing America's Spanish heritage with the arboreal stamp of Anglo dominion.

As the nineteenth century came to a close, the miasma theory fell out of favor, and a public health justification for shade with it. Central Park was a celebrated success, but the business elites who championed it did not have the same appetite for trees on the streets. Olmsted urged cities like New York to commit to curbside canopies that would clean and cool the air in every neighborhood, distributing the park's effect. Yet by the 1880s, the blocks around Central Park were totally built up. The scant street trees that did exist were being whacked by workmen to make room for the hanging wires of telephone and electric poles. The roads of New York, like those across the country, were also being smothered under asphalt, a new material that offered smoother transportation at the same time it ruthlessly deprived underground roots of water and air.

Perhaps the wealthy did not mind these sun-blanched streets because they could escape them. They summered in the country, and some moved to suburbs that offered more verdant environments. But New York's immigrants had no such escape. In 1872, Stephen Smith, a physician who became one of New York's first health commissioners, marshaled massive amounts of mortality data to establish a correlation between the city's high temperatures and the deaths of infants and invalids crammed in hotboxes. It was not miasmas that sickened New Yorkers in the summer, but urban heat. "The day would not seem to be far distant when the resident, especially if he is a laborer, will remain in the city and pursue his work during the summer at the constant risk of his life," the doctor presciently warned.

Smith proposed to cool the air with trees. He calculated that many of the summer's three to five thousand deaths could have been mitigated with shadier neighborhoods. To achieve that goal, Smith convinced the state legislature to pass a law that compelled New York's parks department to forest the streets. But without a budget for planting, the law was ineffective and barely enforced. Undeterred, Smith joined a coterie of civic-minded do-gooders and philanthropists to found the Tree Planting Association. In a furious, two-year burst,

Smith oversaw the shading of the tenements with more than two thousand trees, primarily fast-growing Carolina poplars.

Smith's big idea that the government should provide shade for the urban poor, who were just as deserving of cool air as anyone else, was ahead of its time. Comprehensive planting could have been a standard sanitary measure to improve urban health, no different from clean drinking water, sewer systems, and garbage collection. And it would have made New York a more pleasant place for all. In the early years of the twentieth century, the Tree Planting Association's annual reports opened with photographic illustrations of trees' subtle benefits for urban life. In one picture, a poplar rose from a sidewalk to shade a whole row of open windows, cooling the indoors. In another, a group of children gathered on a stoop to play under an arboreal roof. In yet another, a coachman on a horse rested near the curb under a leafy crown. Shade made homes more healthy, streets more social, and transit more comfortable.

All these basic activities of urban life were not unique to New York. For thousands of years, in ancient Iraq and imperial Rome and the American frontier, city dwellers relied on shade to furnish the comfort needed to travel, walk and talk, work and barter, and come together outdoors. Soon, these activities would move inside. How could Smith have known that just across the East River someone was tinkering with a revolutionary new technology that would change our idea of cooling forever?

Chapter 3

Climate Control: How the United States Turned Its Back on Shade in the Twentieth Century

On July 17, 1902, New York City was melting in a summer swelter. In the midst of a brutal heat wave, the sweaty masses were cramming into the public baths and praying for rain. On the second floor of a Williamsburg printing plant, a twenty-five-year-old engineer from the Buffalo Forge heating company was trying to control the humidity. The steamy city heat was turning magazine pages into blurry messes. The engineer, Willis Carrier, knew that hot air holds more moisture and colder air is drier. He wanted to get that cold air into the printing plant.

But how? Carrier came up with a system to cool air by blowing it across cold water coursing through pipes and then fanning that air onto the factory floor. When this chilled air circulated in the plant, it not only lowered the humidity but by happy accident cooled the space, too. Carrier refined his system further by refrigerating the pipes with a chemical fluid that absorbs heat. He was not the first engineer to try to end the misery of summer heat, but he was the most successful, "a Johnny Icicle planting the seeds of climate control all across America," as one magazine later called him. He conditioned the air in a cotton mill in South Carolina, then a pharmaceutical plant in Detroit and a silk mill in upstate New York. On he went, and in 1915, Carrier left Buffalo Forge to eventually run his own cooling corporation out of a Newark, New Jersey factory.

Industrial products could be manufactured more expertly and more proficiently by transforming humid workshops into chilled environments. Air-conditioning made workers more productive. On the factory floors where it was first installed, it dampened the stifling heat that once caused them to faint. Stuffy banks, shops, and offices no longer closed early in torrid weather. Engineers conducted studies showing how the conditioned air boosted the output of desk jockeys. Global finance was transformed when a massive refrigeration plant and connected system of fans and ducts were installed on the trading floor of the newly redesigned New York Stock Exchange. When the exchange first opened in 1903, the building commission worried the structure would be uninhabitable in the summer due to the sunlight pouring through ninety-six-foot-wide windows. Air-conditioning allowed trading in all seasons.

Air-conditioning changed the American government. In the early twentieth century, Capitol Hill was deserted in the summer. But after the costly installation of air-conditioning in the House of Representatives in 1928, and later the Senate and the Supreme Court, politicians could toil in the factories of democracy all year-round. The White House was next. After an extensive renovation to integrate mechanical cooling, the Truman administration took down the mansion's summer awnings. Conservatives probably scoffed at the wastefulness of a never-ending government spewing hot air at full blast. "The boys don't even know it's sweltering outside," a newspaper columnist said of a summer Congress. "They can orate all through the night and never wipe a perspiring forehead."

Air-conditioning also changed how we relaxed. In the 1920s and 1930s, indoor cooling was a rarity. People flocked to movie theaters and sat through the summer dreck just to refresh themselves in cold air. But after a strong push from Carrier and other corporations, AC was rebranded as "comfort cooling" and came home: first to 12 percent of American households by 1960, then to 36 percent by 1970,

55 percent by 1980, and 88 percent today. Advertisers persuaded women that air-conditioning would save them time—no more cleaning dirt and grime on open windows—and would generate more comfort and leisure, major preoccupations in postwar America. The manufacturers worked hard to take down their enemies: the desire to live outdoors and a traditional summer vacation by the lake, both of which were viewed as sales obstacles. The Crosley Corporation promised consumers that an air-chilled home was as comfortable as a country house or mountain resort. Nature was becoming superfluous—old-fashioned, unneeded, and made redundant by a superior technology that offered, as one ad campaign put it, "the great indoors."

And AC saved lives. Home AC is why Americans are 80 percent less likely to die on a 90-degree day than they were a century ago. In hospitals it is why patients recover faster, surgery is safer, fetal mortality has dropped, and lifesaving drugs are always available. It is why the high temperatures that once threatened the health of pregnant women are no longer so dangerous. Public health officials don't hesitate to prescribe home cooling as preventive care for the elderly and infirm people endangered by the ever-more intense heat waves that broil cities today.

Air-conditioning was a total revolution. But it came at a cost.

This chapter is about how we lost shade to two major changes in our urban environment, one cultural and the other technological. The first part of this chapter is about how Americans fell in love with sunlight. In the twentieth century, urban planners implemented tools like setbacks and height limits that allowed more light to reach the surfaces of dark streets. Architects embraced glass, a luminous material that brought more sun into our homes. But the belief in the beauty and healthiness of solar exposure had an unwanted side effect of making our interiors intolerably hot. The only thing that makes glassy buildings viable is air-conditioning.

The second part is about the consequences of our reliance on this heat-obliterating device. When we moved into conditioned interiors, we lost our connection to the climate outside. Air-conditioning gave us permission to desecrate an urban environment that we no longer relied on for cooling nor traveled on foot. And even worse, air-conditioning began to desecrate the planet, because we must consume huge amounts of energy to create so much artificial chill.

For a few years in the 1950s, some architects tried to control sunlight through design to lessen our dependence on AC. This exciting moment turned out to be exceptionally brief. Now we are stuck with machines that cool us indoors at the cost of hellish conditions outside. Worse, when we all want cooling at the same time, the machines fail us. When our electrical grids black out, millions are at risk of entrapment in uninhabitable heatboxes that can quickly turn deadly.

All this happened when we turned our backs on shade.

When Shade Became an Enemy

In the nineteenth century, office workers had no such expectation of what today's building engineers consider comfort. When the first skyscrapers pierced the air above New York and Chicago in the 1880s, the only method of cooling down was through natural ventilation. The footprints of the ten- and fifteen-story brick towers were T-shaped and U-shaped and H-shaped to create more windows. On hot days, workers in the headquarters of the Woolworth and Equitable companies slid their glass windows open to bring in a breeze from a courtyard or the open air. An electric desk fan might enhance the effect. To move air through the building, tenants opened small transom windows above the doors, which created natural currents.

And to keep the sun's heat out, the tenants relied on what was then a novel device: a retractable canvas awning. This was the AC unit of its day, a small appliance to stick in the window for some seasonal cooling. "The scorching rays of the sun will make you hurry and bustle when real summer weather strikes the town," a Brooklyn manufacturer promised. "Better have your awnings made and put up." Customers could choose from hundreds of window shades in trade catalogs in a range of colors, patterns, and shapes. (A scalloped fringe was a popular choice.) Every year, the awning man came around to apartments and offices to fasten the canvases and their metal frames to window cleats, returning in the fall to collect the devices. Some landlords provided the awning service to tenants free of charge.

As they still do now, desk dwellers had indoor blinds for privacy. The awning outside was for cooling. It was normal to see a clerk leaning out a window of the iconic Flatiron Building, built in 1902, pulling down fabric that had been bunched up like an accordion. Awnings weren't just for offices. Striped awnings were draped over the windows of Victorian mansions and luxurious neoclassical apartment buildings. They shaded the sun-facing windows of the White House and the Florida state capitol, which began taking out the old wood shutters and replacing them with candy cane–striped fabric in the 1890s. And they cooled the upper stories of apartments in Seattle and New Orleans and Barcelona and Paris. So much fabric flapping in the wind, like hundreds of sails on so many stationary ships.

So ubiquitous were these awnings that they even shaded the street. Before the invention of the retractable drop arm, commercial storefronts nailed and hooked their canvases to horse hitching posts and iron railings at the curb, and in doing so covered the sidewalk, too. These commercial awnings didn't have much to do with indoor cooling. Like the Roman porticoes two thousand years before, the awnings allowed retailers to extend the store interior onto the street, where their goods and wares had more public visibility. Not only did the canvas protect the inventory from sun and rain, but it also functioned as a store sign when emblazoned with painted lettering and logos.

When enough of these awnings were extended across the sidewalk, it was natural that a sweaty pedestrian in New York, Philadelphia, or Baltimore might get out of the sunny and dusty road and saunter in the shade for a bit of relief. Practically all of the Bowery was a sidewalk showroom, and in nearby Chatham Square, the canvas gave cover for outdoor saloons and auctions. In other cities, the awnings were not canvas at all but permanent canopies of sheet metal and wooden planks. These ten-foot-tall sidewalk sheds were especially popular in New Orleans, a city that already had a taste for balconies and galleries acquired from the Spanish. Shoppers strolled the riverfront district almost entirely under elevated shelter, free from the drenching sun and soaking rains. The wholesalers and retailers depended on foot traffic, and the shade surely attracted a fair number of walk-in customers.

Innocuous as awnings may have been on office windows, the canopies on sidewalks irritated the emerging class of city planners and civil engineers. The late nineteenth and early twentieth centuries were chaotic times in American cities. Millions of European immigrants who had come in search of a better life were cramming into slums that were overrun and teeming with filth. Worse still, immigrants with nowhere else to go loitered and vended in the shadows of downtowns and business districts. The new municipal professionals wanted to clean the city, not clutter it. Urban congestion, and the related danger of daytime darkness, became their targets.

The situation was dire in the tenements, where entire families were packed into windowless rooms that were thought to be incubating all manner of disease, physical and social. New research from Europe had shown, or so it was thought, that tuberculosis thrived in darkness and the bacteria could be killed with sunlight. Reformers like Robert DeForest, chair of New York's tenement house commission, seized on the findings to bring more health-giving sun into these buildings to stamp out the disease. In 1901, his commission passed a landmark building law to restructure the tenements, carving shafts of light and air into new buildings in the hopes of eradicating the physical and moral abjection. Similar laws followed in Chicago and other cities in an attempt to expose the urban cesspools to the germicidal star. It was not merely physical disease that concerned these reformers. DeForest and others believed dark environments encouraged delinquency and deception. To rid the cities of these urban pathologies, sunlight was again the cure.

Darkness wasn't only a threat to the tired, the poor, and the huddled masses. The shadows of new skyscrapers were also believed to be a menace to the business class. As soon as the towers began rising, so too did concerns about the shade that would plunge streets below into dark and gloomy canyons and shroud the nearby buildings, forcing workers to rely on eye-straining artificial light. One or two skyscrapers wouldn't pose a problem. But if an entire area was to be shaded by twenty-story buildings, a Chicago doctor claimed, "there would hardly be enough sunlight and air to support life." Even the business district could become a breeder of disease, he feared. One leading voice, a Boston architect named William Atkinson, published a book of sun studies and shadow diagrams in 1912, showing how a single skyscraper's winter shadow could fan across the city and deprive residents of sunlight, one hour at a time. "If sunlight is essential for the recovery for the sick, is it not a still more powerful agent in the prevention of disease?" he asked.

So what could be done to make more urban light? One option was to build bigger roads. Proponents of the miasma theory, who believed diseases came from bad air, had long thought a wide road was a healthier one. In the nineteenth century, city builders in London, Paris, and Vienna cut massive new boulevards through the medieval warrens to relieve crowding and bring in more fresh air and sun. (The elderly and infirm reportedly avoided crossing the huge Viennese Ringstrasse because the glare induced "nervous sickness.") American state and local governments, which were not as strong as their European counterparts, acted more gradually. Civil engineers began to broaden the roads to relieve the growing problem of cars, shaving away the sidewalks and the unsanctioned "encroachments" that threatened to force pedestrians back into traffic. Harland Bartholomew, a Newark traffic engineer who rose to national prominence as America's first official municipal planner, had a particular contempt for the pillars, posts, and frames that hoisted the canopies, and urged the removal of "these ugly affairs" on main streets and boulevards around the country. In Poughkeepsie, New York; Allentown, Pennsylvania; and Clarksburg, West Virginia, the "unsightly" sheds and awnings, along with the beggars and peddlers, were removed to furnish the sidewalks with the more palatable infrastructure of electric streetlights.

And urban planners began to solarize the streets through a new legal tool called zoning, which allowed them to control the form of new buildings and thus the size and shape of the shadows they cast. In New York, a powerful real estate lobby refused to assent to limits on ascent. Instead of lopping the tops off new buildings, a municipal building heights commission run by a former congressman named Edward Bassett decreed that above a certain height, the upper floors of new skyscrapers had to be slanted or tiered like wedding cakes to allow more sunrays to strike the streets and the lower stories across the road. The commission called these mandatory sun openings "setbacks," and in 1916, the city adopted them into law. Bassett and the other planners rezoned New York into five height districts. In each one, a new building's form was determined by the width of the street. Midtown, for example, was classified as a "two-times" district, meaning that new towers would start setting back when their height reached twice the width of the street. On a hundred-foot road, a building could be two hundred feet tall before it had to be reshaped to bring in more sun. The suburban fringes of the outer boroughs were zoned one-time districts, a decision that is still visible in their low and horizontal landscapes.

Bassett was drawn into the fray by the construction of the Equitable Building, a hulking, thirty-eight-story office tower that strode an entire city block in downtown Manhattan and cast a thousand-foot winter shadow. Although the problem was said to be the darkness, the loudest opposition came from neighboring landlords who worried the glut of new office space would depress the value of their real estate. For the next thirty years, Bassett later recalled, "the distribution of light and air" in American cities became his main objective. Setbacks went national after then-Secretary of Commerce Herbert Hoover appointed Bassett to help draft what became known as the standard acts, model zoning codes to be adopted by new planning departments in states and cities across the country.

It made sense that in New York, America's densest city, residents would be concerned about access to sunlight. But even more modest cities came to fear a dark future. By 1923, around 220 municipalities had adopted zoning codes of their own. New York's skyscraper tiers morphed into bulk controls on building heights and lot coverage that limited urban development. Row houses, duplexes, and tenements were banned, giving way to new districts of freestanding single-family houses. Milwaukee's land commissioners said it was "common knowledge" that babies wouldn't grow without light and air, which was also critical to the "physical fitness" of adults. In Baltimore, planners ordered sunlight access in new buildings—at least thirty minutes of direct daily exposure to the great disinfectant in every room—claiming that tuberculosis rates went down by half in brighter homes.

In the early 1930s, a planning commission convened by President Hoover turned its attention to suburbia. The commission declared it essential that homes in towns and residential subdivisions receive direct exposure to the low winter sun, an edict that required the provision of clear, unobstructed space that was hard to find in cities. Over time, rules for short, single-family houses and setbacks were written into the design standards of the Federal Housing Administration (FHA), the government agency that backed mortgages. The agency urged neighborhood developers to lay out broad roads that were wider than all but the grandest city boulevards. Eventually, municipalities eager for federal investment adopted these design standards into code. In Los Angeles, where the setback rule was known as the Yard Ordinance, wood-framed stucco houses began to rise, sunny islands on green lawns reachable only by driveways.

To eradicate tuberculosis and create healthier indoor environments, setbacks for light and air would not be enough. Architects began to design new offices and homes that were illuminated by large expanses of glass. The notion of a small square window punched through brick and shaded by an awning was becoming quaint. With the rise of steel frames, it was now possible to clad entire walls in lucent material. Few were more enthusiastic about these new possibilities than the Swiss-French architect Le Corbusier, who declared glass the foundational material of modernism. Le Corbusier and his contemporaries were inspired by sanitariums and antiseptic hospitals where patients convalesced in bright rooms and sunlit porches. As the saying went, tuberculosis was believed to survive for "thirty years in the dark but thirty seconds in the sun." The reformers' faith that sunlight would destroy the pathogen was misplaced. Even as the theory of the sunlight cure was outmoded by the discovery of antibiotics, aesthetic-minded modernists clung to glass all the same.

Because glass is a ruthless heat trapper, these buildings could be unbearable in the summer. Sunlight's visible and infrared waves pour through windows with little resistance and warm the interior surfaces that receive it. Those surfaces then radiate their own longer waves of infrared that cannot pass back through. Gradually, the heat builds up, and rooms bathed in seemingly benign light can turn suffocating. Le Corbusier learned about this greenhouse effect the hard way, first from the overheating caused by the south-facing façade of his Cité de Refuge apartments in Paris, and then as a member of the international team of architects that designed the thirty-nine-story United Nations Secretariat, New York's first glass curtain wall building. The tower was sheathed in three hundred thousand square feet of shimmering skin, far more than any structure in the world when it opened in 1951. Le Corbusier argued for sun-baffling screens called *brise-soleils* to shade the glass, but he was overruled for cost and safety reasons. Snow and ice could collect on the screens, creating maintenance issues and threatening pedestrians. Inside, venetian blinds were installed to block the glare, but they were powerless to fight the heat.

"Something should be said and done about such architecture as this," said Edith Farnsworth, an early victim of glass walls, "or there will be no future for architecture." She spoke from experience. A doctor and art collector who moved among Chicago's sophisticated upper crust, Farnsworth met the German American architect Ludwig Mies van der Rohe at a small dinner party on the city's North Side in 1945 and commissioned him to build her weekend home in rural Plano, Illinois. Mies's vision became the iconic Farnsworth House, a marvel of glass and light that glistened on a clearing in the woods. The house would take its place in the canon of modern architecture, but it caused a great deal of pain to Farnsworth. Construction took years, the final cost was almost ten times the budget, and the house was uninhabitable in the summer. The only solar protection consisted of silk curtains and a single sugar maple, neither of which offered much help. The rest of the forest was too far away to afford shade. A photo from the period shows Farnsworth reclined on a daybed, her hands over her eyes, overcome and exhausted by the heat. She turned on creator and creation, suing Mies and spearheading a campaign against modern architecture.

And just as Farnsworth demanded, something was done about glass architecture. But unlike the ancient architects who oriented their buildings away from the sun and insulated the walls to protect the occupants from heat, or the clerks in the Flatiron Building who opened their windows and pulled down the awnings, Mies did not bother to design a cooler building. Instead, he outsourced the job. "It's up to the engineers to find some way to stop the heat from coming in or going out," he decreed. Indeed, American buildings have relied on massive systems of compressors, ducts, and fans to remain habitable ever since. As thrilling as glass looks, only air-conditioning makes it work as a building material. Architects have never looked back.

American builders used to be climate control experts. In the desert, Indigenous peoples of the Southwest built their homes with adobe to make heat sinks, absorbing the sun all day and reradiating it after dark to temper the overnight chill. Anglos who moved out west adapted their Victorian cottages with high ceilings and vented attics that allowed the hot air to drift out of their homes, and shaded the outside walls with wraparound porches and verandas. In California, Craftsman architects favored the bungalow, a low-slung, horizontal building sheltered under a pitched roof. Some boasted screened-in sleeping porches, where night air offered more comfort than a stuffy interior.

Similar strategies were developed to tame the sultry heat of Louisiana and Texas, where wind, which accelerates the evaporation of sweat, was critical to comfort. In Florida, the Seminoles built open-walled shelters under wide parasol roofs that shed the rain and blocked the sun. Dogtrot, cracker, and shotgun houses were built around tall doorways and long hallways that sucked in cool drafts. Shutters and louvered jalousies drew more air through the windows with only a minimum of sun. In the humid Midwest, cool basements provided an additional summer reprieve. And there was no better patch of shade than a stand of deciduous trees, whose foliage screened the sun in the summer while its bare branches allowed it to shine through in the winter.

But after World War II, the subdivision developers and home-builders backed by the FHA had little interest in constructing houses that couldn't be mass-produced. They bought huge tracts of undeveloped land and offered minor variations on the same assembly-line house to prospective buyers across the country. To do this quickly and cheaply, they had to ignore regional idiosyncrasies of climate. Builders coalesced around two house types to plunk down everywhere: the flimsy, lightweight Cape Cod and especially the California ranch house, boasting sliding glass doors, sealed picture windows, and hardly any insulation in the walls. The lightweight wood frames that supported drywall sheets and low-pitched roofs brought down costs but made for poor thermal barriers.

Because of the setback rules in these new subdivisions, the houses were exposed to the sun on every side, which made cooling essential. This was where air-conditioning excelled. In the early 1950s, with house sales booming, the so-called indoor climate-makers funded model homes in Kansas City, New Orleans, and Corpus Christi, and in the suburbs of New York, Washington, D.C., and Dallas to show how central cooling could be within reach of homebuyers, so long as they were willing to part with the old methods of achieving comfort. A spread in *Newsweek* used information supplied by American Houses Inc. and the Carrier Corporation to do the math. A fixed window saved $20 in building costs compared to an operable window. No louver vent and no attic fan saved another $125 and $250, respectively. And builders could save at least $350 on a house by dispensing with the sleeping porch. "I figured that for the cost of building a Florida room, I could air condition the whole house," one Southern developer boasted, using the local name for the amenity. Of course, these savings really only accrued to the builders. What was not included in their math was the homeowner's expense of actually running the air-conditioning machine, whose costs at the time still outweighed the passive cooling offered by the energy-free alternatives. The FHA further encouraged central cooling when it began to cover the installation costs as part of the mortgage. By 1957, private lenders like the California Federal Savings and Loan Association of Los Angeles required developers to rough in for AC to qualify for a loan.

The same couldn't be said of the trusty old shade tree, which many subdivision builders bulldozed away when they prepared forested land for home construction. Millions of postwar homes were erected without any shade. "We rip down the trees because it's much cheaper to build on cleared land than on wooded property," a builder told *House & Home* in an article about the "economics of trees." "Trees get in the way of trenches, crew traffic from house to house, and the storage of materials." It was "impossible" to spare them during excavation and grading, and even when it wasn't, the "workers and truckmen have no respect for trees," the builder claimed.

No one disputed that a stand of trees added beauty and comfort to a new home. When mature, the trees would also raise the real estate value. But the builders interviewed said they weren't interested because neither the FHA nor the Veterans' Administration "will increase commitments to cover the cost of saving or planting trees. And the buyer won't make a larger down payment to cover their cost." Unlike air-conditioning, underwriters told the magazine, trees were a risk because they could die from pests or storms. *House & Home* continued to acknowledge the drawbacks of a barren site, with another article noting that a "house in a treeless tract must handle about twice as big a heat load as the same house in the woods." But it made no financial sense for builders to absorb the costs of trees. That made "air cooling" all the more important to their buyers.

Air-conditioning didn't just make new designs endemic. It also rendered old ones obsolete, and whole ways of life with them. Consider the porch, which was long the place where American families came together to cool down. This is a transitional place, like its predecessor the Roman portico, that straddles indoors and out. It's where we get out of the sun and rain while getting fresh air. The porch is where we see what's happening in the neighborhood from the safety of our home and, with a nod or a wave, become a part of the scene. Numerous writers commemorate the folksy and neighborly quality of the porch and rhapsodize about the custom of coming together in the place where "the soul can rest." It's where we relax, where we share stories, meals, and songs. Porches are even sites of romance, the intimate setting for a courtship—that is, until someone inside turns on the light and ends the evening encounter.

But air-conditioning ended the allure of the porch, or in some cases engulfed it, as families decided to screen in the room, glass it up, and turn it into another climate-controlled interior. On a summer afternoon, take a moment to listen to the sound of your neighborhood. What you're likely hearing is the whir of compressors. In 1979, the *Time* magazine columnist Frank Trippett anticipated this sad state of affairs when he wrote that air-conditioning had already "seduced" families into retreat behind closed doors and windows. He lamented the loss of a society whose "open casual folkways were an appealing hallmark of a sweatier America." He moaned that instead of chatting with neighbors on a summer evening, most Americans were watching TV. These days, we are buried in our technological cocoons. By bringing people inside, air-conditioning robbed shade of one of its most critical historical functions as a place where people gather.

The Path Not Taken

For about a decade after World War II, air-conditioning's centrality in American life was not assured. Americans endured fuel shortages and restrictions on coal and oil. Even in peacetime, the threat of long-term scarcity loomed. It was conceivable that the United States would have to return to war and continue to fight Communist foes. Were rations next? To preserve energy for future conflicts, government researchers seized upon two decades' worth of weather data that defense agencies had assiduously collected, and fashioned it into practical advice for architects and building engineers to integrate "climate" into house design. Suburban housing was the fastest-growing business sector in the country. Done right, these new homes could even be more livable and less expensive than air-conditioned ticky-tacky.

"It is doubly important now that climate assume its proper position as one of our natural resources," read the foreword to *Climate and Architecture,* a slim 1951 guide published by the research arm of the Housing and Home Finance Agency (HHFA). The urgency was not unwarranted. By that time, millions of Americans had already moved into suburban tract housing. A national addiction to heating and cooling was looming, thanks in no small part to the financial support offered by the FHA, one of the HHFA's subagencies. Two years earlier, Congress appropriated a small sum to the HHFA for housing research to find some alternatives to a growing dependence on AC. Joseph Orendorff, the agency's research director, knew he had a tough row to hoe. Nevertheless, he believed homebuilders could be persuaded to design and erect sustainable housing through careful studies and sufficient guidance.

Orendorff believed environmental concerns would soon be "as indispensable to the design of buildings as structural engineering is today." Solar orientation and sun shading were deemed critical, and to perfect the art Orendorff turned to Victor and Aladar Olgyay, two Hungarian architects who approached climate architecture as a science. At the time, the twin brothers were working at the Massachusetts Institute of Technology, where they had fastidiously translated the military's weather data into an array of diagrams on which they could pinpoint man's "comfort zone." Although the concept of a comfort zone was first devised by Willis Carrier, the air-conditioning inventor, the brothers were determined to achieve it first by architecture and design, then engineering.

The Olgyays were accomplished architects in their native Hungary. Back home, the brothers had designed more than forty buildings, including a squat Budapest apartment known as the Reverse House for the balconies that faced away from the street. They designed it this way not so the tenants wouldn't have to deal with the noise and dust of the city—although that was surely a benefit—but so the apartments would receive the maximum amount of winter warmth from a backyard that faced the sun. When the brothers moved to the United States in 1947, they became fixated on the scorching heat and wet-blanket humidity they had not experienced back home. Instead of harnessing the power of the sun to cut down on heating bills, the Olgyays used their talents in service of the opposite aim: not to capture sunlight, but to make shade.

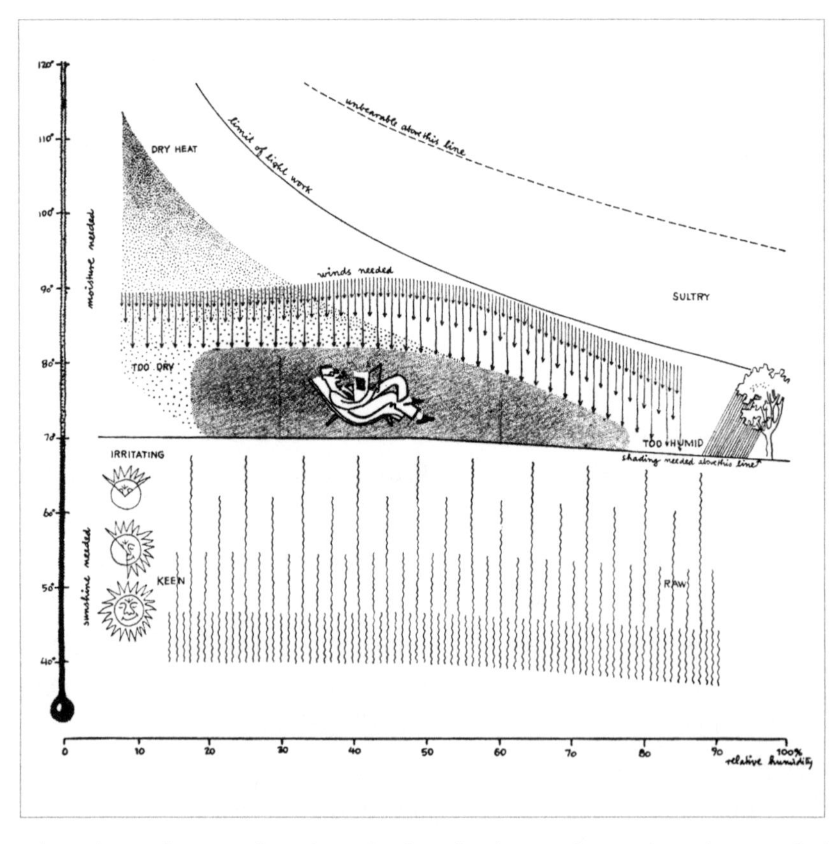

The reclining figure in the Olgyay brothers' bioclimatic chart is located in an ideal "comfort zone" determined by air, moisture, wind, and shade.

In a 1954 pamphlet, the brothers laid out their method for designing with the climate. The first step was to map the local meteorological conditions onto a modified version of Carrier's two-dimensional graph that the Olgyays renamed the "bioclimatic chart." The brothers correctly asserted that human comfort was not determined by air temperature and humidity alone, as Carrier believed. The effects of sun and wind were critically important as well. In a schematic version of the chart, their revised comfort zone was presented as the conditions under which a bespectacled man in full business dress, reclining on a chaise and smoking a pipe, would not break a sweat. Although comfort is subjective, the brothers believed, as many did at the time, that there was a quantifiable range that could apply to everyone in a given climatic region. They believed that ideal comfort could often be achieved without the active mechanical cooling of air-conditioning.

The Olgyays plotted air temperature on the chart's vertical axis and relative humidity on the horizontal axis. For a typical American, thermal neutrality could be located in the area between 70 and 82 degrees and between 30 and 65 percent humidity. The Olgyays believed 70 degrees was a crucial threshold. This was their designated shading line. As air temperature dropped below 70 degrees, people needed sunshine to stay comfortable. Above the shading line, the opposite was true: Exposure was an irritant. Shade by itself could furnish comfort in warm temperatures. However, the beneficial effects began to diminish at 82 degrees. Above that temperature, the addition of a breeze was necessary to restore the feeling of comfort, especially as humidity increased.

Outside the comfort zone, the Olgyays drew two long, descending lines, starting at around 115 degrees. The first curve dropped to 95 degrees in moderate humidity and down to 80 degrees in sticky air. This was the "limit of light work." To achieve comfort while cleaning or vacuuming, more moisture was needed in dry heat, and more wind was needed in humid heat. The second curve indicated extreme conditions. It fell to 110 degrees in moderate humidity and tapered down to 95 degrees as the air was progressively saturated with moisture. The weather was "unbearable above this line." It was the upper limit of human health. In such conditions, chores were the least of someone's problems. Between these two lines, comfort could not be achieved by the natural means of climate and architecture alone, and mechanical cooling was necessary. But AC should be used only after all other possibilities are exhausted. For comfort, architects should start with shade.

How could architects create that shade? In a 1957 book called *Solar Control and Shading Devices,* the Olgyays urged them to review regional climate data to identify the days of the year when shade was the prerequisite for comfort. Next, they instructed architects to diagram the sun's path in the sky on those days, determining the angles at which sunrays would pass through a window and fall on the surfaces of a room. After calculating the heating effects, architects could then design a solar control solution in the form of a shading device to attach outside the window. The shape of the device was determined by the shape of the shadows needed to be cast inside. For reference, the Olgyays included a handful of photographs of shading devices on buildings around the world, illustrating how the problem of solar gain could be mitigated with taste and style. Today, most of us probably read these window louvers, roof overhangs, and egg crate–shaped façades as the groovy and geometric decorations of mid-century modernism. But this solar hardware was critically functional, the exterior expression of interior comfort. For a moment, the future wasn't bright. It was shady.

Climate-responsive architecture was the preserve of not just anxious federal officials and nerdy professors but the lifestyle-conscious middle class as well. In 1949, the interior decorating magazine *House Beautiful* began to explore the possibilities. Then edited by Elizabeth Gordon, the magazine espoused the conviction that utility and taste went hand in hand. Houses were not meant to be gaudy but to fulfill basic human needs. Recent oil shortages had prompted Gordon, like other Americans, to reflect on the need for conservation, and she believed climate's influence on domestic living was crucial to prog-

Mid-twentieth-century architects used "shading devices" to save energy and avoid overheating, such as this fixed metal awning on a Clayton, Missouri, department store.

ress. Energy efficiency was just the kind of better living project that fell under *House Beautiful*'s umbrella. Gordon believed that existing houses provided shelter but not control, and she set out to educate her readers, providing them with strategies, methods, and design schemes to better their lives and the world.

Gordon sought both a sensible riposte to the Cape Cods that were strewn across the country and an American response to the glassy expanses of Mies van der Rohe and the so-called International Style. The magazine's "Climate Control Project" probed readers to ask how the elements could be captured for comfort. At what angle did the sun enter the living room? Where did the wind come from? And how might a well-placed overhang or some additional landscaping make life inside more pleasant? Prototypes appeared in *House Beautiful* as Pace Setter Houses, attempting to inspire more than half a million readers to reconsider what a dream home could be. The spreads were premised on a commonsense approach that would sound familiar to environmentalists today: Spend more up front on a comfortable and sustainable house for lower heating and cooling costs in the future.

Gordon's ambitions were not unrealistic. In the 1940s, shelter magazines like *House Beautiful* regularly sponsored model homes and had a major influence on American architecture. *Ladies' Home Journal* promoted affordable Victory Houses and Frank Lloyd Wright's utopian Usonian houses to a broad audience. In 1945, *Arts & Architecture* began sponsoring the Case Study Houses that are today icons of California mid-century modernism. *House Beautiful* didn't publish floor plans for Pace Setter Houses. Instead, the articles and accompanying photographs taught readers the concepts and lingo they would need to instruct architects to build climate-sensitive homes, like site planning, solar orientation, ventilation, and shading. Compared to the Olgyays' more scientific methods, the approach of *House Beautiful* was more down-to-earth. There were few charts and diagrams. Instead, readers learned climate-control tricks.

In 1949, the *Architectural Forum* editor Henry Wright explained to readers simple techniques to "put a harness on the sun." He urged homebuilders in temperate zones to locate more windows on the south side to take advantage of the diminished winter sun. To block overhead summer sun, he recommended horizontal projections. The sun was harder to control on the east and west exposures, especially at sunrise and sunset, and Wright told readers to be strategic about glass there. But he cautioned that it would not be enough to locate windows in the right places. Wright pointed out that summer and winter temperatures lagged about a month behind their sunpaths, which meant that architectural solutions like overhangs were sometimes ineffective. August could bring an infernal combination of low sun and high heat. For this reason, the "natural screening" offered by a heavy growth of deciduous trees was an ideal complement to "structural sun-controls."

But just as Wright was teaching readers about sunpaths and the Olgyays were perfecting their method of bioclimatic design, technology and economics rendered their architecture obsolete. The Olgyays' processes for achieving sustainable comfort with a minimum of energy were derided by other architects as too dense to interpret, by university colleagues as ridiculous, and by human physiologists as bewildering. In some ways, the brothers were ahead of their time. Computer software now easily sorts through weather data to spit out sunpaths and shadow diagrams, which makes building for shade theoretically more accessible. But their obsession with a thermally perfect interior was out of step with the new formal and design possibilities that were entrancing their colleagues. The brothers believed architecture was science, but others believed science only served to frustrate an architect's expressive voice. A voice, of course, that only became more expressive with the assistance of ubiquitous air-conditioning.

At a 1952 government-sponsored building research conference in Texas, the attendees came to the conclusion that architecture alone could not produce clean and sweatless living. Shade could reduce the amount of heat a room absorbed, but it had little effect on the humidity that stifled an occupant's ability to bear it. (A notable exception were the louvers and window shutters that also induce airflow.) The best that shade could ensure for someone in Austin was "tolerable discomfort" and that wasn't going to cut it. Likewise, the aspirations of Gordon and *House Beautiful* editors to show how ordinary homes could maintain harmony with nature proved to be out of touch. The essential premise of thinking about a house as a system for managing climate was too complicated for most suburban homeowners moving into prefab housing. Many of them saw their home not as a natural air conditioner but as a way to build wealth and enter the middle class. It proved difficult to buck the suburban stereotype of crushing conformity. At the height of the postwar building boom, construction on more than a million houses started every year, and most of them were mass-designed to conform to the developer's street grid, not the seasonal paths of the sun. Anyone who chose to turn their house away from the street and subvert the norms risked being seen as pretentious or unneighborly, maybe even a Communist.

In 1954, the congressional appropriations for the HHFA research arm that promoted the Olgyays' methods ran out and the unit folded. Environmental design returned to the fringe. The agency continued to fund developers to build houses, not execute sophisticated climate solutions. Sensitive landscapes, shady courtyards, breezy hallways: all these things take up room that could instead go toward more housing. The possibilities of a different way of living—a thriftier one, a sweatier one, or even a less productive one—were eclipsed by the oil and gas that again flowed through the United States. It wasn't just the cheap energy. Glass got better, too. In the 1970s and 1980s, gold and silver coatings that subtly deflected infrared radiation while admitting the visible spectra began to flood the market. These low-emissivity windows are much cheaper than architectural shading devices, and they cut out enough heat to get much of the job done. And as for whatever's left, there's air-conditioning.

"There's basically a conceptual understanding of why you would need shading, but that's kind of where it ended," lamented Seattle architect Michael Eliason about his American architectural education. Eliason learned the art of solar control in Germany, where manufacturers still install the awnings, shutters, and blinds that Americans once relied on. When we spoke in 2021, he had just designed a school with a massive wall of south-facing glass that was protected by external shades. But when the budget ran over, the shades were among the first things the design team cut as part of a loathsome process known to architects as value engineering. Obviously, there was a cheaper and simpler way to prevent the building from overheating. "You just oversize your air conditioner, and poof—you've solved the problem," Eliason said. "It's the default mode of thinking."

The Air-Conditioned Nightmare

Before air-conditioning, people came together to beat the heat. New Yorkers escaped stifling apartments by taking in the air on their doorsteps, a custom that was observed even on Manhattan's wealthiest streets. On sweltering nights, they slept on roofs and fire escapes. Others camped at the beach. In a classic essay in *The New Yorker,* the playwright Arthur Miller remembered when his family survived a summer cauldron by sleeping on the grass in Central Park, among hundreds of other families. People in Boston, Chicago, and Lincoln, Nebraska, did much the same. These days, it's illegal.

Air-conditioning is part of the gradual movement away from outdoor life. Plugging in AC allows us to disconnect from nature and the communities it fosters. It encourages privatism. In 2018, Rowan Moore, a British architecture critic for *The Guardian,* visited Houston, one of many Southern cities that was, in a sense, created by the air conditioner. Before AC, 24 percent of Americans lived in the Sunbelt; by the year 2000, it was 40 percent; by the time Moore arrived, it was closer to 50. He was awed by the desolation of the streets and the apparent preference of Houstonians to spend whole days and weeks indoors, moving from "your air-conditioned house to your air-conditioned garage and then in your air-conditioned car to parking garages, malls and workplaces which are all, also, air-conditioned."

Like other foreigners visiting America, he wondered where the people were. In Houston, he found them inside—specifically, in a network of tunnels and skywalks, six miles long, that took them from hotels to food courts to banks without once setting foot on a public street. What saddened Moore about Houston was not just an aesthetic crime but the fact that he could not find a true public space that was open to everyone. Instead, he found air-conditioned interiors that guests paid to enter. (Beyond the tunnels, public libraries are a notable exception to the rule, but those in Houston, as in most U.S. cities, have seen their hours dwindle as budgets shrink.)

Air-conditioning doesn't just exclude heat and humidity. It also excludes people. In places like Houston's tunnels, security officers bother visitors who aren't there to shop or work—the homeless, rabble-rousers, protesters, bored loiterers, people with nowhere to go, or people who look the wrong way. The tunnels exclude people with physical impairments, because some were built before the passage of the Americans with Disabilities Act and aren't elevator accessible. Moore observed what the writer Henry Miller in the 1940s dubbed "the air-conditioned nightmare": a country that was divorced from nature and devoid of culture, but whose residents didn't seem to mind all that much because they were comfortable.

So many hours spent behind climate-controlled glass also makes us physically intolerant of summer heat. When air-conditioning began killing us softly with treated air, architects like Frank Lloyd Wright worried about an unhealthy addiction to artificial chill. He was right to worry. Studies have found that our reliance on AC makes us more sensitive to heat. People who spend more time in air-conditioned spaces are more likely to feel overheated and are more vulnerable to heat illness than those who are more often exposed to higher temperatures. In part, that's because they live more sedentary lifestyles—driving in an air-conditioned vehicle instead of walking a hot street—and haven't developed the fitness to withstand the environmental stress. But it's also true that continuous cooling degrades the body's ability to tolerate deviations from the norm. Americans have adapted to air-conditioning and weakened their ability to survive without it.

Fortunately, our addiction to air-conditioning is largely reversible. That's according to Stan Cox, author of *Losing Our Cool: Uncomfortable Truths About Our Air-Conditioned World*. Early in the book, Cox introduces a couple who moved from New Jersey to Phoenix and decided to wing it without AC. The couple's perception of heat became more flexible. As the weather warmed, they felt comfortable for longer, acclimating to higher and higher temperatures. That the human body can make this adaptation has been borne out by U.S. Army studies. After five days of working out in high temperatures for about two hours, subjects' bodies showed significant tolerance of heat, and after ten days they acclimated almost completely. Their natural cooling mechanism of sweating kicked in faster and more effectively. Presumably, the same training that prepared them to fight on a sweltering battlefield would also help them live in an apartment without an air conditioner. Cox emphasizes that AC has a lifesaving role in extreme situations. But he suggests that rampant use may be one reason why Chicago's heat wave in 1995 killed more people than two hotter, more intense ones in the 1930s. Many Americans have grown used to cooling and forgotten how to live with heat.

In the summer, when demand for AC soars, electrical grids around the world are pushed to the limit, forcing what one analyst called the " 'oh shit' moments" that precede blackouts. A huge one in the U.S. Northeast in 2003 caused approximately ninety deaths, from people who suffered accidental carbon monoxide poisoning from generators to those whose cardiovascular, respiratory, and other health conditions may have been triggered by the sudden onset of heat. Elderly residents of high-rise apartments were stranded without power and potable water. Firefighters rescued hundreds from elevators. Subway riders were trapped in the dark, scared and stressed out. In New York, on a day when the air temperature only reached 91 degrees, the lack of AC risked human health. That kind of summer swelter is already common and will be more frequent in the future. The apparent inability of our power systems to handle it should concern us.

Brian Stone, Jr., an urban climate expert at Georgia Tech, studied what would happen in some of America's largest cities if they blacked out during a heat wave. His conclusions were grim: hundreds of thousands of people at risk of heatstroke and other illnesses. In Phoenix, a rapidly growing desert metropolis that's one of America's fastest-warming cities, easily half the city would need emergency hospital care, and about fifteen thousand people would die, a situation as deadly as the most intense hurricane we can imagine. "You do remember, for example, that Puerto Rico lost electricity for six months?" he asked me. "There's no reason to think that can't happen in the mainland U.S. Whether it would be six months, I doubt it. But could we have two weeks in Phoenix without AC? Absolutely. We should assume it's going to happen."

When Phoenix officials were presented with the results of Stone's study, their response was that people would simply leave the city. Residents would abandon their homes and flee in their vehicles—that is, if they owned one and had fuel in the tank. Gas station pumps don't work when there's no power. And Stone found Phoenix didn't have an evacuation plan. There was no designated escape route. There weren't enough cooling centers—air-conditioned rooms in libraries, schools, malls, and convention centers—to shelter thousands, much less millions, of potentially stranded people. And there weren't even mandatory backup power generators to keep those cool refuges up and running.

Stone's team also found that people who didn't have air-conditioning at home would probably fare better in these doomsday scenarios. They had "greater climate resilience" because they were more likely to take precautions to deal with the heat—simple moves like drinking more water, drawing curtains, turning off the lights, and putting foil over windows. Living without AC forces you to get creative. And it prepares you to deal with the worst.

Besides posing a risk to our health, air conditioners also exacerbate the heat problem they're supposed to solve. Air conditioners don't make spaces cool by making heat disappear; they move hot air somewhere else. What happens when you crank AC in a hot room? A compressor increases a chemical refrigerant's pressure and temperature, sending it to the condenser coil, where the refrigerant turns into a liquid. This liquid then passes into the evaporator coil, and a fan blows indoor air across this coil, allowing the refrigerant to absorb the heat. The now-cooled air is distributed throughout the house, while the heated refrigerant gas is passed back to the condenser coil, which releases it outside. This is why we feel a muggy blast when we walk past window units chugging away in the summer. Air-conditioning is also why New York's low-ceilinged subway stations reach stifling temperatures of 120 degrees. The stations trap all the heat that the coolers are expelling from air-conditioned train cars.

On the other side of the planet, a research team in Singapore used infrared cameras to record air conditioners installed on building façades belching hot plumes into the air on a sultry night. The units were all stacked atop one another, so the heat expelled from one unit outside got sucked right into the unit above it, and on and on it went, forcing them to work ever harder to cool the rooms. Computer simulations from Paris and Phoenix estimate that all the waste heat from ACs adds about 2 to 3 degrees of warming to nighttime air, making them a significant contributor to urban heat. One person's cooling is someone else's thermal garbage.

The problem isn't just that ACs cool our interiors by ejecting hot air, but how they do so. First, there's the issue of refrigerants. Air conditioners use chemical compounds called hydrofluorocarbons (HFCs) to absorb heat. These fluids leak out of air conditioners, and they're extremely potent greenhouse gases. The most abundant HFCs are 1,300 times more effective at trapping heat in the atmosphere than carbon dioxide. Governments around the world have pledged to cut these emissions, and companies are looking for climate-friendlier alternatives, but federal scientists estimate that leaks only account for about 40 percent of the global warming caused by air conditioners.

The bigger issue is just how much energy they need to remove heat and push it outside. In a typical American home, air-conditioning consumes the most electricity. These appliances use twice as much energy as our refrigerators, four times more than our clothes dryers, twelve times what our ovens use, and twenty-eight times as much as our dishwashers. Americans use more energy for cooling than the billion-plus people of Africa use for everything. And this wanton consumption is not sustainable, because it's powered by burning fossil fuels. Cox estimates that air-conditioning and refrigeration already account for more than 10 percent of all greenhouse gas emissions, and that share could double in the next twenty-five years. It's not an overstatement to say the pollution we generate for cooling is burning the planet.

Theoretically, the United States could eliminate our cooling emissions if we get on a clean grid. But that ambitious achievement would not address the growth in cooling happening outside North America, on continents where large numbers of people are rising out of poverty and staking their claim to comfort. The electricity needed to serve the new global middle class is almost certainly going to come from dirty stuff first. New coal-fired plants will have to come online. By 2050, the International Energy Agency predicts, somewhere north of five billion AC units, four billion more than we have today, will be chugging along in high-rise towers around the world, producing two billion more tons of CO_2 every year. If the past is prologue, then any gains from smarter and more efficient machines will be completely overwhelmed by the massive rise in demand.

And that's the awful irony of it all. Between the hot air they expel into our cities, and the global warming caused by fossil fuels burned to power them, our reliance on air conditioners is fated to grow stronger. We had the chance to take a different path. We could have spurned the aesthetics of glass skyscrapers and the economics of one-size-fits-all homes to live more sustainably. We could have kept heat at bay instead of trying to obliterate it, and in the process, make it stronger. We could have embraced shade. Instead, we doubled down on AC.

Part II

Desperate for Shade in the Twenty-First Century

Chapter 4

Surviving the Sun: For Agricultural Workers, Finding Shade Is a Matter of Life and Death

Maria Isabel Vasquez Jimenez was born into poverty in a Mixtec village in the mountains of Oaxaca. As a teenager, she dreamed of becoming a hairstylist and opening a salon. She knew she would not be able to make this dream a reality by earning $4 a day selling tamales. Over her mother's protests, the seventeen-year-old contacted a coyote to smuggle her to Lodi, California, to live with her uncle and earn fifteen times that much as a farmworker and twenty times if she worked hard. She followed her fiancé, a skinny nineteen-year-old named Florentino Bautista, who was working in almond orchards and saving money to buy her a gold engagement ring. They planned to stay in America for three years before returning home.

Florentino was employed by Merced Farm Labor, a contractor that assembled crews to work the massive agricultural fields of California's Central Valley, where half of America's fruits, nuts, and vegetables are grown. Not long after Maria arrived in Lodi, she joined Florentino. On May 14, 2008, a van picked up the teenagers and drove them to a white wine vineyard in nearby Farmington, where Maria was assigned to tie grapevines on trellises for $8 an hour. They arrived just after sunrise. Low rays of morning light raked through the waist-high rows of plants spanning hundreds of feet. Maria and Florentino got to work.

Four hours later, a thirsty Maria looked up, squinting into the sun. She had been sweating for hours, and her back ached and legs wobbled. Her face was streaked with sweat and dirt. The air temperature was approaching 95 degrees, and the farm's overnight dew had wafted into a humid mist. Scanning the vineyard for water, she saw that someone had stationed two five-gallon coolers at the corners of the rows. But rather than waste ten minutes walking there, she chose to continue working. It was only her third day on the job, and if she didn't meet her quota, the foreman might not bring her back. Besides, she didn't see anyone else breaking for water. Maria and Florentino worked through lunch, too.

At around 3:40 P.M., something was wrong. Maria hadn't sipped any water all day. Her hands and feet were swollen, and her heart was pounding. She had a throbbing headache and could not remember where she was. When she tried to speak, her words slurred. She closed her eyes to dim the glare of the white-hot field. Then she collapsed. Maria fell to the dirt and Florentino ran over to cradle her ninety-pound frame. Maria's clothes were wet, but her face was pale and dry. Florentino urged Maria to wake up. He did not know that she had lost control of her own body. No amount of water would help her now.

Someone alerted the foreman, who wandered into the rows to see what was happening. "It's just a regular fainting," he said, hovering over the teens. In fact, it was heatstroke. When the brain senses heat, it vasodilates, bringing blood to the skin's surface. But when this goes on for hours, and the internal organs are starved of blood flow, they begin to break down. The gut weakens and leaks toxins into the bloodstream. Those toxins threaten white blood cells, and they launch a defensive response by creating blood clots, which in turn leads to organ failure. This is the beginning of an irreversible domino effect. One part of the human body fails, then another, and another.

Three years before, in 2005, the California Division of Occupational Safety and Health (Cal/OSHA) passed a landmark rule that guaranteed workers like Maria a legal right to shade, water, and rest breaks to prevent heat episodes like this. But the laws on the books were not the laws on the fields. Maria and other laborers still had to work when there wasn't any shelter from the sun. Where foremen had not been trained to recognize the onset of symptoms that snowball into heatstroke, the lack of enforcement proved deadly. That summer, Maria was one of at least six California farmworkers to perish in the sun. Two of them were veteran field hands who succumbed to triple-digit temperatures as they loaded grapes into trucks. Dozens more fell ill.

The foreman ordered Florentino to leave Maria in the van until the shift ended. Florentino wet her bandanna and pressed it against her forehead. Twenty minutes later, when the crew piled in to be driven home, the foreman saw that Maria's condition had not improved and ordered the driver to fetch some rubbing alcohol for her at a drugstore. But the *raitero* sensed this was not a mere fever and decided instead to take her to a clinic. The foreman called the driver, who handed the phone to Florentino. "Remember two things," the boss told him. Number one, this did not happen at work. Tell the doctors Maria fell ill when she was exercising. And number two, *no tiene papeles*. You don't have papers. The implicit threat was that the foreman could get Florentino deported if he did not do as he was told.

The clinicians could tell that Maria was not sick from exercise. She was comatose. They called an ambulance, and at 5:15 P.M. she was admitted to Lodi Memorial Hospital, where ER doctors recorded a core temperature of 108 degrees, far beyond the limits of the human body. Her organs were cooked. Maria stayed in the hospital for two days as doctors tried to revive her. It was there that Florentino learned Maria was two months pregnant and had suffered a miscarriage. The official cause of her death was visceral multi-organ hyperthermic injury due to job-related heat, but California farmworkers have another name for what happened to Maria: *se quemó*. She burned up.

Maria's uncle contacted the United Farm Workers (UFW), the venerable California union, to seek some help. It would not be long before her death created a public outrage. Less than two weeks later, Governor Arnold Schwarzenegger came to her funeral in Lodi, where he held Florentino's tiny shoulders and listened as the boy remembered his bride-to-be. "This land gives us a lot of opportunities, but gave her death, and we have to make sure this doesn't happen again," Schwarzenegger said. He comforted Florentino and told him Maria's death should have been prevented. In fact, he had previously stated that heat illness was "very, very dangerous" and professed a commitment to "take care" of farmworkers.

Marc Schenker agrees. A former University of California public health scientist and work safety expert, Schenker has studied farm conditions since the 1980s. Back then, pesticide exposure was the main health threat in the fields, but due in part to his advocacy it's been more than eight years since a farmworker was poisoned to death. As more workers like Maria died from heat, he turned his attention to the issue. People who worked under a fierce sun for eight or ten hours straight turned up at hospitals with their organs on fire. Their bodies were warped, stiff as a board with arms and fists balled up. The survivors were permanently disabled by liver failure, brain damage, and paralysis. When it comes to heat, there's no deadlier place to work than a farm. Field hands and crop pickers are nearly three times as likely to die from heat than construction workers, and twenty times more likely than the workforce as a whole.

Schenker worries that as temperatures become more extreme, cases like Maria's will become more common. Climate scientists say workers like Maria are "climate canaries," because what happened to her could be an early warning of danger for the rest of humanity. "These are not diseases or deaths that increase with age," Schenker said of heat fatalities. "They're not chronic diseases. They're not genetic disorders. They're caused by environmental factors that we know how to prevent." The tragedy is that we don't do it.

A Critical Tool for Thermoregulation

The physiologist Michel Cabanac, who studied thermal alliesthesia, knew the sun's impact is not only skin-deep. It gets in our heads. People who spend too much time in the sun feel strange and disoriented. Sometimes they forget what they're doing. Some say that sluggishness is the default mode in hot climes and sunshine makes us slackers. It's an observation as old as the sun. The Enlightenment philosopher Montesquieu, who believed climate determined the character of societies, disparaged deserts where "the excess of heat enervates the body, and renders men so slothful and dispirited." Others have painted a picture more insouciant, like Mick Jagger, who sang that "the sunshine bores the daylights out of me." Cognitive impairment is an unfortunate side effect of our body's natural response to heat. Cerebral blood flow is reduced, which in turn slows our thoughts by depriving our brain of the oxygen it needs to function.

Sunshine appears to trigger these cognitive effects in ways that hot air alone does not. For decades, physiologists have studied the sun's specific influence on the brain. A few years ago, Japanese researchers strapped sensors to high school baseball players and monitored the vital signs as they practiced in the morning sun. After three hours, the athletes' hearts were pounding, and their core temperatures rose to 102 degrees, indicating overheating and possible heat illness. But the athletes didn't feel hot or fatigued. This suggests that sunlight dulls the signals to the brain that communicate danger. More recently, Canadian and European physiologists studied the motor skills of young adults exposed to simulated sunshine. The researchers had test subjects ride stationary bicycles in front of iodide lamps for an hour and then tested their alertness and memory. As their skin warmed, they were slower to react and made more mistakes. The mental fatigue was less apparent when the cyclists pedaled in darkness.

In another experiment, test subjects sat in front of computer monitors, with similar solar lamps pointed at the backs of their necks. Besides the expected discomfort, the scientists also observed that hand-eye coordination worsened. It seemed that radiation could penetrate the skin tissues and even the skull to cook the cerebral cortex. As human physiologists know, one of the brain's first responses to heat stress is to behaviorally thermoregulate. But seeking shade was not an option for those test subjects. According to Leonidas Ioannou, a physiologist who studied the cyclists, their brains instead sent out a different message to the rest of the body: slow down.

Outdoor laborers often have neither luxury. Farmworkers and construction workers in Qatar, the Persian Gulf country known for the fatal combination of burning heat and stifling humidity, have rarely benefited from shade or rest breaks. (Following the international controversy that arose from the deaths of thousands of workers building the facilities for the 2022 World Cup, the country implemented these basic protections.) There, a team of physiologists ran a study comparing two groups of workers: one that labored in 105-degree air in the shade, and the other in cooler 91-degree air in full sun. The scientists used a metric called the wet-bulb globe temperature (WBGT) to compute the total heat stress caused by the four meteorological variables of air temperature, humidity, wind speed, and solar radiation. Even though the WBGT was about the same for both groups, the workers in the sun seemed to be hurting more, despite the cooler air. After a day's work, the sun-drenched workers were twice as likely to report headaches and muscle pain, three times as likely to feel tired, and four times as likely to feel lightheaded. These were the symptoms that Maria almost certainly experienced as she was pushed beyond her limits in a California vineyard.

Some basic physiology explains how humans like Maria overheat. When we work outside, we gain heat internally, as a byproduct of metabolic activity. The harder we work, the more heat we create. Strenuous labor and intense exercise can turn our hearts into furnaces and raise heat production by ten or twenty times compared to a body at rest. And we gain heat externally, from the world around us. We exchange heat with the environment through four main processes. The first process is convection, through the air. The second process is conduction, through direct contact with another surface. The third process is radiation, through electromagnetic waves. Outside, the main source of radiant heat is the sun, but we can also gain heat from surfaces the sun has warmed. And the fourth process is evaporation, when heat is lost through sweat. We do not gain heat through evaporation, only lose it.

According to Robert Brown, a landscape architecture professor and microclimate specialist at Texas A&M University, of the three sources of environmental heat gain, radiation is the most influential and the easiest to control. This may seem counterintuitive for the simple reason that most of us probably think about heat in terms of air temperature, not sun exposure. Although a humid 90-degree day may feel hot, in fact, we do not gain heat from air until it is warmer than our skin, which has an average temperature of around 91 degrees. As the difference between our skin temperature and air temperature narrows, we lose less heat through convection. The same is true of humidity: When the air is saturated with moisture, we lose less heat through evaporation. On such a day, we really only gain environmental heat through radiation.

And on warmer days, when air temperature does exceed skin temperature and we gain heat through convection, Brown has calculated that radiant heat remains the dominant environmental factor. As an example, he computed the heat gained by a human body standing in the sun on a 105-degree day in Phoenix. The heat gained through radiation and convection were both problems, but according to Brown, the sun had transferred ten times as much heat to the human as the air did. The mere act of standing in the sun was the rough equivalent, from an energetic perspective, of a moderate workout or a brisk walk uphill. It's often said that the wall of hot and dry air in Phoenix feels like an oven. By comparison, the sun is a microwave, shooting heat energy straight into our flesh. Much of that microwave effect could go away, Brown pointed out, by simply slipping into the shadow of a nearby building.

When radiant heat gain is added to the significant heat generated by hard labor, the burden can be too much to bear. Humid conditions create what physiologists call uncompensable heat stress, which is a situation where the body is overwhelmed by the heat and unable to offload it. The first sign of this struggle is excessive sweating, which is the body's most powerful attempt to shed the heat. Excessive sweating dehydrates us, and that in turn can cause muscle cramps, fatigue, and nausea. These are the symptoms of heat illness. The most serious illness is heatstroke. This occurs when the body has depleted its water and can no longer cool down through sweat. Inside, as the body temperature rises, the internal organs conk out, and release toxins into the bloodstream. Heatstroke victims are manic or delirious, because their brain is failing. Eventually, the core temperature rises to 104 degrees, and the victim dies.

But heatstroke is not the only way heat kills. People with chronic health conditions, whose hearts cannot handle the physical demands of sweating and blood pumping, could suffer cardiac arrest. And people with physical disabilities have their own vulnerabilities. "For some of us, waiting for the bus could put us in the hospital, or in some cases, cost us our life, and that's not hyperbole," said Jennifer Longdon, who used to represent a Phoenix district in the Arizona House of Representatives. Two decades ago, Longdon's spine was shredded in a drive-by shooting, leaving her paralyzed below the collarbone and permanently disrupting her skin's ability to communicate with the hypothalamus and trigger autonomic thermoregulation. Longdon does not shiver in the cold or sweat in the sun. Instead, the heat simply builds in her body. After twenty minutes of sun, she feels weak and lightheaded. Sights and sounds grow distant, like she's on the other side of a tunnel. She behaviorally thermoregulates by seeking shade, or "sweats" by pouring bottled water all over her bare arms and shoulders and waiting for it to evaporate. "The idea that someone could be outside, and die within forty-five minutes sounds nearly impossible," she told me, "until you start recognizing that there are folks like me that don't have a functioning autonomic system."

Because shade reduces the heat from radiation, it has surprisingly large health benefits. It is critically important to the elderly, whose bodies are more sensitive to heat. As we age, our nerves weaken, which messes with thermal communication. Our sweat glands atrophy, which inhibits our ability to lose heat through evaporation. And we become more likely to take medications that indirectly put us at risk for heat injury, like dehydrating diuretics or antihistamines. The less heat the elderly must dissipate, the better, which makes thermal behavior like seeking shade all the more important. The Arizona State University heat scientist Jennifer Vanos, one of Brown's former students, led an international team that modeled the limits of livability in extreme weather. According to their simulations, on a moderately humid 95-degree day or a dry 105-degree day such as the one Brown examined, an elderly woman can safely hang out in the backyard or stroll with friends, so long as she is hydrated and in shade the entire time. If she tried to do the same in the sun, she could die in three hours.

As heat mounts, it becomes more dangerous. Without a break, our body can continue an inexorable march to collapse. Experiments on outdoor athletes show how shade can offer an opportunity to recuperate. In a study of stationary cyclists in Copenhagen, the late Danish physiologist Bodil Nielsen and her colleagues observed the physiological impact of a thirty-minute period where the cyclists were shaded under a reflective umbrella. Even as they continued to pedal, the cyclists' heart rates fell by eight to ten beats per minute, compared to cycling in the heat. Their oxygen consumption fell, which meant they were breathing easier. Their sweat rate slowed by about 20 percent. Even after they returned to the sun, their core temperature continued to tick down for fifteen minutes. These findings about shade's effect on stamina have a profound implication for other people who cannot slow down outside, like farmworkers. If provided with shade, they can be safer, and even more productive. An international study found that farmworkers in Florence and Guangzhou worked faster when they were sheltered from the sun during sweltering harvests. The researchers urged farm managers to invest in mobile structures, like large umbrellas and wheeled gazebos, that would allow harvesters to work in shade instead of just resting in it. Similarly, researchers have discovered that seed packers in Indonesian rainforests are more productive than those in logged clearings. Using environmental heat stress monitors and oral thermometers, the researchers found that shaded workers are 39 percent less likely to overheat than those in the sun, and not coincidentally, take fewer rest breaks to cool down.

These lessons about the health benefits of shade are also driving interventions in Central America, where sugarcane cutters die by the thousands from chronic kidney disease caused by dehydration. Normally, kidneys fail as people age, but in these rural villages, men in their thirties and forties are spending their final days hooked to dialysis machines. There, researchers are monitoring the impacts of shade in the fields. In the 2010s, after cutters in El Salvador began using wheeled tents as they worked, they reported fewer headaches, cramps, and nauseous episodes. In Nicaragua, where cutters carried four-post Saran mesh canopies, kidney injuries dropped by 70 percent.

There is growing evidence that chronic kidney disease is also rampant in the produce combines of California. In 2014, Marc Schenker and other University of California researchers found that Central Valley farmworkers were showing elevated levels of creatinine, an indicator of kidney dysfunction. Many of these farmworkers are piece-rate earners, meaning they are paid by volume and leave money on the table with every bathroom break. Like Maria, they had the legal right to rest in the shade but did not believe they could afford to exercise it.

"A young person will come to that field and say he can do a hundred buckets," said Amadeo Sumano, a forty-four-year-old farmworker. "I will kill myself working today, and sacrifice tomorrow's performance, because I'm working hard today," he said of these young hotheads. I met Sumano on a drowsy August afternoon in Santa Paula, California, where field hands sprinted through farm rows hauling twenty-five-pound buckets of green bell peppers. The air was thick with a spicy aroma. Sumano, who knew better than to expose his skin to the sun, wore two long-sleeved shirts, blue jeans, and a wide-brimmed straw hat. At the end of the field, I noticed a twelve-foot brown tarp pitched over two benches and a wooden table, or "shade trailer," so named because it is hitched to a pickup truck and wheeled onsite with the porta-potties and mobile sinks. This was the mandatory shade structure that was compelled by California law. Officially, foremen stagger breaks to ensure everyone can rest in the meager shade, but advocates say that doesn't actually happen. Many workers shelter in their cars instead. The foremen aren't worried about the regulations, because the state has fewer than two hundred safety inspectors to police an estimated seventy thousand farms and ranches.

As for the workers themselves, they simply accept the consequences of laboring in the sun. "One of the clinics in town gave me information that said farmworkers have high rates of kidney disease," Sumano said. "It's a fact. It's what we live with." He shrugged. "But so what? I can't do anything about that." Sumano barely earned minimum wage and couldn't afford a blood test, much less comprehensive healthcare. He feared heat not because it could hurt him but because it could cripple the harvest and cause financial ruin for his family of four. In thirty years, I told him, there would be many more days of 95-degree heat. Perhaps another month's worth in Santa Paula every summer. "Wow, *pesado,*" said Roman Pinal, a UFW organizer who translated our conversation. Pinal asked Sumano what climate change meant to a farmworker. Sumano was unmoved. "We're already carrying that load, so we'll just carry it longer," he said. "We will put up with it. We will do it." As long as there's money to be made in the fields, farmworkers will suffer under the sun. But shade could provide some relief.

The Promise of Mandatory Shade

After Maria Isabel Vasquez Jimenez died in 2008, a veteran farmworker named Maria de Jesus Bautista (no relation to Florentino) came home from a Riverside County vineyard complaining of a headache and nausea. She died in a hospital two weeks later. Ramiro Carrillo was picking nectarines in a field in Fresno County when he told the boss he wasn't feeling good. Although it was only 10:00 A.M., he was already suffering from heatstroke, and that evening he died on his front porch. In 2009, UFW and the surviving family members, along with other farmworkers who feared for their safety in the fields, sued California for failing to protect Jimenez, Bautista, Carrillo, and hundreds more laid up in sickbeds and emergency rooms. The suit argued that California had the knowledge and power to prevent heat illnesses. Other employers had managed to protect workers in grueling environments. Why should farmworkers be any different?

The union urged California to follow the lead of the institution that had thought about heat longer and harder than almost any other: the U.S. Army. For decades, the military had combated heat illnesses among soldiers not merely by providing water, rest, and shade but by ordering them to use these resources. A key figure in the development of the army's heat protocols was a physiologist named Edward Adolph. In 1937, the University of Rochester scientist was invited to study heat illness in Boulder City, Nevada. The desert town was founded only a few years earlier to house the builders of the Hoover Dam, the landmark project to harness the power of the Colorado River and supercharge the growth of the American West. Adolph's fellow researchers became his test subjects. He observed that during strenuous exercise, the scientists did not feel thirsty, even though they lost gallons of water to sweat. Like a farmworker nearing the end of a row, the scientists remained focused on the task at hand, even as it accelerated their own physical decline. Adolph coined this aversion to water "voluntary dehydration." The same biological process that regulated a person's temperature also caused dehydration and left them too clueless to know it.

During World War II, the military commissioned Adolph to study human performance at an army air base in Blythe, California, with the aim of preparing soldiers for combat in the deserts of North Africa. Between 1942 and 1945, Adolph and colleagues conducted hundreds of tests on the troops, marching them through the shadows of bunkers and down asphalt roads and sandy trails for miles on end. On a hot, sunny day, Adolph noticed that soldiers sweat about a quart of water every hour. He calculated that one cup was stimulated by direct sunlight. A second cup was drained by hot air and ground radiation. And the third and fourth cups were lost to metabolic heat. The solutions seemed to be clear: Get out of the sun, take breaks, and stay hydrated. And yet again, Adolph noticed, his subjects did not feel tired or thirsty. It was not enough to make water and shade available, he concluded. The recruits had to be forced to drink and rest.

Adolph's research became critical to the U.S. Army when it was discovered that a surge of recruits had died in boot camp. Between 1942 and 1944, 198 troops died of heatstroke in training, a death rate that was four times higher than in active service overseas. In 1952, during a brutally humid summer at the tough Marine Corps camp at Parris Island, South Carolina, six hundred recruits were laid low with heat illness and exhaustion. What was happening to trainees wasn't sustainable, and Adolph's research laid the groundwork for three key innovations that are standard in military institutions to this day.

First, a flag policy to communicate the severity of the weather conditions on training bases. Based on the WBGT, drill sergeants and commanders signal limits on physical activity by semaphore. On days with mild weather, commanders fly a green flag, indicating that it is fine to run strenuous drills so long as seasoned troops drink water and take a brief rest every hour. On hot days, they run a yellow flag, which requires more water and longer breaks. A red flag indicates dangerous heat and the need to rest in shade every half hour. And a black flag is a shutdown.

The second innovation was the formulation of the WBGT in the 1950s. Unlike the National Weather Service's (NWS) heat index, which combines the effects of air temperature and humidity, WBGT also accounts for the sun and wind. The heat index is important because it indicates our body's declining ability to cool down and the danger of extreme temperatures. But it assumes someone is resting in the shade, not exercising under the sun. (Although the NWS notes that direct sunlight can raise heat index values by 15 degrees, such information does not appear in the weather apps on our phones.) WBGT sensors measure radiant heat with a thermometer tucked inside a small black globe. The temperature of this globe is weighted twice as heavily as the so-called dry-bulb temperature of the air, reflecting the sun's more significant, though often overlooked, contribution to heat stress. A wet-bulb thermometer that measures humidity is weighted heaviest of all. Because the WBGT assesses the comprehensive threat, it is standard in the military, recommended for work safety, and even favored by high school football coaches, who have a responsibility to protect their charges from heat.

The third innovation is what the military calls "acclimatization." Over two weeks, recruits are gradually exposed to more heat, which builds up their body's ability to withstand it. For the first five days, new trainees can exercise in the sun for only two hours a day. Gradually, exposure ramps up in duration and intensity. Over time, the body learns how to respond to heat: The heart doesn't work as hard, the core temperatures decrease, and more sweat is produced. In ten days, the process is done. The change isn't permanent, though, and repeated exposure is required to maintain the adaptation.

Together, the flag policy, WBGT, and acclimatization worked. The incidence of heat illness on Parris Island fell from 53 victims per 10,000 trainees in 1953 to 4.3 by 1960. By 1980, the protocols were adopted across the military. In 1990, as Operation Desert Storm loomed, military doctors and scientists prepared a sophisticated heat strain model that computed strict work and rest schedules specifically for soldiers in the Middle East. On sunny days, whether troops were on base or fighting in the desert, they were supposed to rest under whatever shade they could contrive, like an open canvas tent or a camo net strung to the side of a tank. As the temperature rose, more shady rest was mandated. On a 91-degree day, soldiers were supposed to rest every forty minutes. In triple-digit temperatures, rest was ordered every half hour. Although the protocols may not have always been followed in the heat of battle, casualties were nevertheless low, army researchers found, because of the commanders' discipline.

The protocols originated by the military spread to the world of work. The National Institute for Occupational Safety and Health (NIOSH), a division of the Centers for Disease Control, leaned on army innovations to develop its own "criteria for a recommended standard" to protect workers from "hot environments." By 1986, workplace safety experts, sports medicine specialists, and at least seven countries, including the USSR, had all adopted heat guidelines and procedures. Maybe, NIOSH suggested, U.S. workers needed those protections, too.

On rare occasions, NIOSH has been able to act on that ideal. Work safety experts point to the cleanup of the Deepwater Horizon oil spill as a paradigmatic example of successful heat management. After the BP oil rig exploded in 2010, the agency coordinated with the U.S. Department of Labor to ensure that a hastily assembled workforce of forty-seven thousand beach cleaners would be protected like the soldiers of Desert Storm. Every morning, the crews threw shade across the shoreline, dotting the beaches in pop-up tents. BP was required to provide shade for every worker at rest. Even the vessels were covered, in the form of plastic sun cloths stretched across shrimp boats and oil skimmers. Because of the sweltering Gulf Coast heat, the crews were sometimes only allowed to work for ten or twenty minutes before they had to cool off in a tent. No one involved in the cleanup suffered heatstroke—the silver lining of an otherwise devastating environmental catastrophe.

Experts have taught us how to stay healthy in the heat. We can acclimate to higher temperatures. And we can survive with the provision of water, rest, and shade. But unfortunately, as Adolph discovered, our bodies aren't great at telling us when we need these things. And sometimes, the pressures of the workplace make it hard for us to prioritize our own physical needs. In the language of the military, we must be commanded to do it. And just as the army has mandated these protections for soldiers, labor advocates in California hope to do the same for farmworkers toiling in the sun.

The Fight to Bring Shade into the Fields

Clear skies and an afternoon high of 102 degrees was just what Dean Florez had in mind. It was July 28, 2005, three years before Maria Isabel Vasquez Jimenez would perish in a vineyard. Florez, a state senator from Shafter, a farm town in California's southern Central Valley, stood on the edge of a cotton field, barking into a bullhorn.

That summer, like every other in recent memory, migrants were passing out in the fields, sacrifices at the altar of California's powerhouse agricultural economy. Florez's colleague, a Los Angeles–area state representative named Judy Chu, had introduced a bill to protect California's millions of outdoor workers from the heat. But the bill had flatlined and was close to death on the floor of the state assembly. Over the objections of farm owners and Republicans who thought the heat rules excessive, burdensome, and crippling to their bottom lines, Florez was trying to bring the bill back to life. "Many of you will never be in a field again," he said, leaning against the tailgate of a Ford F-150. Lawmakers, aides, farmers, and reporters found their seats on five-gallon paint buckets upside down in the dirt. Accustomed to air-conditioned offices, they found no respite from the heat but the occasional gust of wind. By the end of the day, Florez hoped his opponents might have a better appreciation of what it's like to work under the sun.

At the time, California workers had been demanding heat protections for years. In 1984, Los Angeles librarians laboring in stifling stacks petitioned the state's workplace safety agency, the California Division of Occupational Safety and Health (Cal/OSHA), to adopt existing heat guidelines as state regulations. (Indoor heat can be dangerous, too.) In 1990, the sister of a tomato picker who was crushed to death while escaping the sun under a tractor trailer demanded safe, reliable shade structures on farm fields. But it wasn't until 1999, when Cal/OSHA calculated that employers were suffering from lost productivity, that the agency convened an advisory committee to hash out some rules. Anne Katten, a work safety specialist with the California Rural Legal Assistance Foundation and a member of the advisory committee, drafted regulations to protect workers from heat. Those rules collected dust until 2005, when Chu tried to make them law in the California legislature.

Chu represented the L.A. suburbs, not farm country. Because of that, she believed she was in a unique position to help rural laborers. "In some ways, it's easier for an urban person to push hard against agriculture," she told me from her office in Washington, D.C., where she is now a congresswoman. Of the powerful farmers who control one of the state's largest industries, she said, "I'm not affected by them, in terms of their support or their possibly vitriolic attitudes." It was these farmers and their lobbyists who managed to sideline her 2005 bill, arguing that an existing state program already protected workers from injuries and illnesses. But Chu explained that the program was ineffective. "We found twenty-one preventable heat-related deaths from 1996 to 2004, well after this so-called program was put in place," she said. "And those were only the documented cases." In California, farmers must protect their animals from the weather or face a felony charge and jail time. Chu thought their workers deserved the same humane treatment.

Although in 2005 Democrats controlled the California legislature, the Republican minority united in opposition to Chu's bill, and Governor Arnold Schwarzenegger threatened a veto. But after a then-record heat wave, that summer would turn out to be a deadly one in California's fields. July was the height of pain: "A month that brought shame to the Valley," said *The Sacramento Bee*. Though Schwarzenegger opposed Chu's bill, he was sympathetic to the farmworkers' cause. Before he was a politician, Schwarzenegger was an action movie star and had suffered his own bout of heat illness on the set of the movie *Predator*. He had lain in bed for six days, suffering through headaches, muscle cramps, and even torrents of vomit and diarrhea. He knew how serious heat illness could be. Moreover, he was a Kennedy by marriage, and his famous in-laws had supported the UFW since the days of Cesar Chavez. He wanted to continue the custom, but he would do so on his own terms.

Florez played a key role in the process. The state senator came from generations of workers on Shafter vineyards, a tradition his father ended by graduating from high school. "The lessons that I remember, growing up, were always based on farmworker values, as I call them," Florez recalled in an interview from his home in Pasadena. "Work hard. Keep your head down. Don't complain. And be thankful to have a job." Not long after Florez was elected to the state assembly, a van crashed into a tractor trailer on a county road in August

1999 and killed thirteen farmworkers on board. He was surprised to learn that it was legal in California to crowd them into vehicles without seatbelts. "That's just what farmwork is," he said. "You get in a van, and you pack yourself in, and you're happy to get out to the field. If people die, it's just part of the harvest."

For almost a hundred years, field hands have been exempt from federal laws that guarantee a minimum wage and basic safety protections. They must get by on their own, without much help from the government. Ray Florez, Dean's father, relied on his smarts. Like Amadeo Sumano, he wore long sleeves and layers of clothes to protect his skin. He didn't drink alcohol on the weekend if the weather forecast called for heat on Monday. And when he needed a break at work, he crawled under a truck to snatch some shade. "That was his solution," Dean told me. "Find anything." Unlike the crushed tomato picker, he lived to tell the tale.

His family history taught Florez that "people believe that farmworkers are simply implements." Farm owners, known in the agriculture industry as growers, did not bother to ensure the labor had basic necessities like seatbelts, protection from pesticides, and shade. In 2005, as Chu's bill foundered, Florez pitched the growers in his district on pop-up canopies like the ones that his daughter's swim team used at the pool, and that his family gathered under for front yard barbecues. He imagined crew bosses popping them open and workers scooching the posts down the rows as they picked the fruit. "I felt they would have shade, pretty much, for a good majority of the day," Florez explained. But to the farm owners, "that was crazy." He reeled off the objections. The shades were a nuisance. The work would slow down. They would lose money. He scoffed. "The arguments were based on, 'we've just never done it this way and we're not going to do it this way.'"

Undeterred, Florez brought his idea to Schwarzenegger. "We can't have pop-ups in the fields," he recalled the governor saying, during a meeting in a spacious cigar tent that Schwarzenegger erected in the capitol courtyard to get around an indoor smoking ban. "There's plenty of shade," an aide added. "People can go to their trucks and sit opposite of the sun. They can use the grape leaves as a partial shade." Florez was incensed. "This isn't Napa Valley," he thought. "These trellises are four feet tall. You're forcing a farmworker to sit on their butt, get on their knees, and crawl underneath a dusty, pesticidal bush. That isn't shade. That's crap." That's when it hit him. Was it possible that Schwarzenegger and his aides hadn't ever been on a California farm field during a harvest? That they had no idea what it was like to toil under a crushing sun? In a burst of inspiration, Florez decided to move the conversation outside.

"I think it's critical that we get out of our air-conditioned Capitol chambers and back whatever decisions we make with a real-life understanding of what these workers endure," Florez said in a press release announcing his outdoor event, which he dubbed a "Meeting in the Sun." The meeting was officially billed as a public conversation about the prevention of heat-related deaths and illnesses, but it was really an opportunity for the union and their allies to garner support for the bill. For two hours, around 130 lawmakers, aides, farmers, union officials, student activists, and journalists gathered to drink warm water, fan themselves with paper, and listen to workers describe the horrific circumstances of their loved ones' deaths: A forty-two-year-old veteran farmworker collapsed in a field of bell peppers and died an hour later. Another worker who asked for a break near the end of the shift was found lifeless among the vines the next day. That very month, three workers died in the fields, and a fourth, a twenty-four-year-old picker, was lying in a Bakersfield hospital bed, having been airlifted from a remote tomato field. In the rows, said a grape harvester, "it feels like I'm drowning." Florez and his assembly colleague Chu listened. The "sound-bite virtuoso," as the local paper called Florez, was having trouble putting one word in front of another. The heat was getting to him.

The gambit worked. The next day, Schwarzenegger's office called Florez. The governor declared the situation in the fields an emergency and wanted a rule hammered out. Florez spent the weekend on marathon conference calls with Schwarzenegger and his political confidants and allies, who included a farm owner and an agribusiness lobbyist. On August 2, 2005, the governor held a press conference in Sacramento to announce the broad strokes of emergency rules to prevent more suffering. From now on, he declared, every outdoor worker would have water, rest, and shade. His workplace safety department would hammer out the details. Florez stood behind him, gritting his teeth. Over the objections of the growers, he got his pop-up canopies, but they came with a caveat. Workers had to ask for them—and if the army's studies were reliable, they wouldn't.

Schwarzenegger's emergency decree was promising, but it was not the last word on heat illness. Although he mandated water, rest, and shade, the governor did not say how they would be provided. This opened an opportunity for interest groups to participate in public rulemaking. On one side were doctors, work safety experts, and labor advocates pushing to implement quantitative rules similar to the U.S. Army's, which they believed to be the gold standard. On the other were growers, farm owners, and business associations who believed strict protections would hurt their bottom lines. After a decade of meetings and negotiations, California's regulators arrived at rules to protect outdoor workers. Nevertheless, some experts contend the rules are not strong enough, for the simple reason that outdoor laborers still perish in the sun.

When Cal/OSHA opened the rule in 2005, one of the first tasks was to arrive at a definition of shade. This had never been done by military heat researchers like Edward Adolph. Business representatives lined up to inquire about the legality of the many opaque objects that were already strewn about their workplaces. Was a metal toolshed with an open door considered shade? Was a hard hat with a big brim considered shade? What about the mesh scrims that diffused light on movie sets? Or a tarp dragged over grapevine trellises? It seemed they would do anything to avoid buying the pop-up canopies that Florez fought for. The state's attorneys decided none of those objects counted as shade, though trees did. Eventually, they came up with a simple definition: Shade is made by an object, natural or artificial, that blocks direct sunlight and cools a human body. A parked car that has been baking in the sun and is suffocatingly hot does not make shade, even though it blocks sunlight. But a car that is idling with the air-conditioning on does, because it cools the body.

The rules of provision proved harder to hash out. Work safety experts, also known as industrial hygienists, argued that outdoor workers should drink one quart of water per hour. (As Adolph knew, that is how much is lost to hard work under the sun.) Watercoolers should be close by to encourage workers to drink; Anne Katten suggested no farther than a thirty-second walk away. She also pushed for dedicated shade structures to cover every outdoor laborer during their rests and lunch breaks, located within a one-minute walk of their workstations. Physicians argued for mandatory rest breaks to prevent heat illness, because as army commanders know, an overheated worker may not be aware they are in danger. Howard Spielman, one of California's leading industrial hygienists, suggested that foremen and other labor managers use the WBGT to determine rest schedules, just as the U.S. Army did.

More interested in efficiency than in health, agribusiness resisted expert opinion. Farm owners, labor contractors, and business representatives from the California Chamber of Commerce said the suggested protocols would be inconvenient, impracticable, and expensive. They claimed that providing shade for every worker on a crew could cost $35,800, plus thousands more in annual replacement costs. They claimed that too many pop-up canopies in the rows would cause traffic jams and congestion in the fields. They claimed that too many breaks would force the farms to leave ripe fruit on the vine. They even claimed the rules were condescending. "What's next?" complained Robert Roy, a lawyer from an agribusiness coalition. "They're going to tell us when we have to go to the bathroom?"

California's heat protection rules emerged from this history of conflict and compromise. Frequent drinking is not mandated but "encouraged." Instead of specifying where coolers must be located, the regulations instruct labor managers to place them "as close as practicable" to workers. Instead of the mandatory rest commanded in the military, laborers are allowed to take a break "when they feel the need to do so to protect themselves," contradicting researchers' findings that overheated people often do not realize they are in danger. Thanks in part to the advocacy of civil rights lawyers, there are now high-heat procedures that kick in at 95 degrees. But instead of a fifteen-minute break every hour as some safety experts suggest, the rules mandate a ten-minute break every two hours. Instead of shade in every row, the rules require "one or more areas with shade" for the entire field. And although the rules specify that there should be enough shade for every worker on a rest break, that is not the same as enough shade for every worker, period. These protections are vastly better than none at all. But they are a product of political expediency, not careful scientific research, and for that reason workers like Maria Isabel Vasquez Jimenez continue to die.

To some, California's heat rules are a model of success. Inspectors visit about three thousand worksites a year, and about two-thirds have plans available for water, rest, and shade. Reported heat injuries have decreased by about 30 percent. And even as temperatures continue to rise, the number of heat deaths has fallen to about two every year. Protecting workers has not appeared to hurt the farmers' bottom line; in fact, the state's farm economy keeps growing. Over the years, other states adopted their own heat rules, and in 2024, President Joe Biden's Labor Department looked to California when it got the ball rolling on a national standard.

Yet heat scientists and workplace safety experts still say these heat regulations aren't good enough, because they consider every heat illness and death to be preventable. At least seventeen workers died between 2014 and 2024, and that's just the official tally. California doesn't record deaths that occur after an overnight hospital stay, or others that happen after work or on small farms. This makes it hard to quantify heat's true impact on the state's estimated 800,000 farmworkers. Between 2001 and 2018, California workers filed around 360,000 compensation claims for injuries triggered by high temperatures, or roughly 20,000 per year. Heatstroke, heat exhaustion, fainting, ladder falls, machine wounds, and other accidents were 6 to 9 percent higher on days above 90 degrees and 10 to 15 percent higher on days above 100 degrees. In cities, it's usually the elderly who fall prey to heat, but in the fields the injury rates are highest among young men.

In summer 2021, then the hottest in California history, I drove from L.A. to the Central Valley, coming down over a rugged canyon to cruise a long, flat highway through farm country. Thirteen years earlier, when Maria died, it would have been a different sight. I would have passed fewer almond trees, now more numerous because they can be harvested with machines that save labor costs. I would not have seen so many beach umbrellas at the ends of the farm rows or shade trailers on the sides of the road. And the skies would have been brighter. Two wildfires raged within sight and many more were burning up and down the Central Valley. In the distance, smoke piled into the atmosphere, blotting out the sun. The sky looked like a dark bruise.

A few miles past McFarland, I got off the interstate and turned onto a county road. Row after row of grapevines were bagged in white plastic that protected the fruits from the elements. I pulled onto a dirt patch near a canal and waited for my guide, a smiling outreach coordinator for the UFW Foundation. Together we ambled down a gravel path carrying N95 masks and hand sanitizer to distribute to the crews already deep in the acres. The sun was shining and *ranchera* music was blasting. It didn't feel like one of America's deadliest places to work. Yet when I asked workers about heat, their mood changed.

Stefanie Diaz, a twenty-eight-year-old Bakersfield woman, was swaddled in flannel and multiple bandannas. After ten years in the fields, the sun was now wearing her down. On hot days, when her legs began to tremble, she steadied herself against a metal table at the end of the row where she packed grapes. The table is covered by a corrugated roof and a dangling blue tarp, but I was told that shade was for the fruit. Recently, on another job in a pepper field, she had been unable to keep up with a harvest machine and thought she might faint. Sometimes, when workers like her pass out in the field, they are simply left behind. "Every year, it's two or three degrees higher," added Primitivo Cruz, a heavy-eyed forty-year-old picker snipping shriveled fruits off a bunch. He pulled at one of his shirts, fanning his chest. "I sweat more. I'm tired." He knew people on other farms who died, and knew the heat could get him, too. Still, he didn't take breaks because he earned more money with every carton of grapes he filled. Like Diaz, he made sure to drink water, but water cannot ward off heat illness on its own.

When the foreman called for lunch, there wasn't enough shade for everybody. The mandatory pop-up tents were occupied by the boss sitting at his table and the workers who brought sodas and chicharrones to sell to their colleagues. Almost everyone else ate lunch under grapevines in the field, or in their cars where they cranked AC. Twenty years after shade became law, such shade shortages are still typical on California farms. Labor contractors know they can get away with breaking the rules because the state's workplace safety agency is seriously understaffed. There are fewer than two hundred inspectors policing a state of eighteen million workers, and only thirty-five of those inspectors speak Spanish. That's less than one inspector for every ninety thousand workers. It can sometimes take a week or two to initiate an inspection after a complaint comes in. By then the crews—and the contractor who hired them and is legally responsible for their shade—have moved on.

The state still does not require shade for every worker and instead suggests a workaround of rotating breaks. Although this sounds like a practical solution, Juanita Perez, a community worker at the California Rural Legal Assistance Foundation, said she's "never seen it happen." She would know. A daughter of farmworkers, Perez spends the harvest season in her car on the sides of county roads, watching farm operations through binoculars. She rarely sees umbrellas and tents for the whole crew, or clean water or restrooms nearby. Perez files hundreds of complaints to Cal/OSHA every year. Most are ignored, dismissed, or rejected. She is dismayed but not deterred by the state's inaction. "I kind of feel like I'm the only one out there," she said, watching out for the farmworkers.

"We've had some inspectors tell us, 'There's trees, right? There's grapevines, and they're casting shade,'" she said. "We push back against that, because we want a structured shade where people can sit comfortably and actually rest, and not feel like, 'I'm going to sit on the ground real quick, before I start my row.'" She sighed. Vines are inadequate for shade, she explains, because they do not separate workers from their relentless, steaming conditions. "The employer creates the environment where it's like, 'You're not supposed to rest, you're a body, and we need these grapes, and we need these raisins, and we need these out,'" she continued. "A lot of my uncles and aunts are still working in these fields, and when I see someone without a shade, it's almost like, you're treating my family like animals."

So long as there are impoverished workers who are willing to risk their lives to make ends meet, injuries will continue to pile up. And so long as the supply of those workers remains virtually endless, employers will allow them to take those risks. For his part, Marc Schenker believes the piece-rate system that drives workers to death should be banned, but he knows that's unlikely to happen. There are other things California can do. It can make the labor laws conform even more closely to what heat scientists and safety experts say is needed. For example, Oregon requires more paid rest breaks as the temperature climbs: at least twenty minutes every hour when the heat index is 95 degrees, rising to forty minutes every hour at 105 degrees. So far, the rules seem to be working. In 2021, Oregon saw 109 heat-related deaths. In 2022, when the rules kicked in, the toll fell to 22 such deaths, and in 2023, just 8 people died, despite many days of triple-digit heat.

And Dean Florez, who left his term-limited office in 2010 and now sits on the California Air Resources Board, still wants more shade. "Let's not give up on the idea that canopies can roll with farmworkers," he said. Go out to the coast, he told me, and see how the farmers protect their precious strawberries and raspberries, growing them under white plastic stretched over steel hoops. The sheeting protects the berries from sunscald: spasms of white drupelets, curdling on the sunny side of the flesh. "We care so much about the coverage of that plant, and that product, that we've got it down to a science, what the shade does for them," he continued. "But no one gives a rat's ass about the farmworker having that same type of protection, because they're not valuable."

Florez worked hard to implement California's regulations mandating water, rest, and shade. And yet he still sees the suffering. "I will drive home from Pasadena to Shafter, where my parents live, through the 99, and see the fields, and all these workers out there," he said. "I don't see any kind of tent, or rolling tent, or any sort of cover that allows them shade. I see an umbrella here and there, at the end of the row, where they're bagging grapes. I will see little shade. I will see minor improvements. And believe me, it doesn't feel good." We spent a few minutes spinning out the possibilities. In California, tractors pull packing stations through celery and bell pepper fields. It wouldn't be hard to drag a roof through the fields, he said. It pained him to see that after all those years of fighting for shade, his efforts appeared to have culminated in protection for grapes. "You think you've done something, but you know that job is really not done, and there's all these ways out of it," he said. "There's a lot of work still to be done. A lot of work."

Chapter 5

The Shady Divide: In Sunny L.A., Keeping Cool Separates the Haves from the Have-Nots

Debbie Stephens-Browder woke up and peeled herself off an imitation leather interior. A rail-thin woman with an infectious smile, the sixty-six-year-old retired schoolteacher became homeless in 2017. After she lost her childhood home in Compton, she laid her head down in women's shelters, first for a spell in San Francisco before returning to South Los Angeles. In 2020, when the Covid-19 pandemic hit, she caught the virus and the shelters kicked her out. Where was she supposed to go? She had already exhausted her savings at a Motel 6 and could not ask her family to put her up. So like many Angelenos down on their luck, she started sleeping in the back seat of her car, a beat-up Mercedes-Benz sedan with a fickle ignition.

Stephens-Browder used to love sunny days, but she dreaded them when she became unhoused. A daily search for a place to beat the heat became harder when libraries closed during the pandemic, which deprived her of free air-conditioning. Cars are decent shelters from the rain, but the glass windows and dashboard make them cauldrons in the sun. The suffocating Benz pushed the lifelong asthmatic to the edge. To survive the heat of South L.A. streets, she needed shade.

And for a time, she had it. Stephens-Browder summered under the thick crown of a majestic ficus at the Watts Civic Center, but when the towering botanical structure was felled for urban development, she had to get out of the sun. And in April 2021, when I met her, she was still searching for shade. At one point, she parked overnight in a gated lot that a local nonprofit reserved for Angelenos like her, who lived in their cars. Besides a few rows of swaying palms, there wasn't much cover, yet the conditions were tolerable, in part because Stephens-Browder was somewhere else in the light of day. She had become a community organizer for TreePeople, a stalwart environmental group that planted trees in Watts, a neighborhood where she had familial roots. Trees fight the heat that worsened her life on the streets and even put it at risk. She moved up to the driver's seat and turned over the ignition.

"When I talk about the heat in Watts, I really know all about it," she explained. "Because I can't do anything to escape it."

You can see L.A.'s shady divide from outer space. On one side of the sprawling city are wealthy enclaves built around golf courses and tucked into leafy canyons. Mansions nestle under the canopies of majestic oaks and sycamores, and shoppers stroll beneath their manicured crowns. On the other side are neighborhoods where vast concrete expanses are open to the sun: playgrounds, parking lots, bare sidewalks, and wide roads. Bus riders bake on unsheltered streets and homeless people seek refuge under highway overpasses. Some came to L.A. to seek their sunny American dream only to suffer relentless heat.

Why do L.A.'s wealthy neighborhoods have ample shade and poor neighborhoods have hardly any at all? One reason is the technical challenge of growing trees in a semiarid city built for cars. Another is the political challenge of mustering support for shade's maintenance and provision. And then there are the hostile attitudes about public space. For decades, L.A. police have discouraged trees because their shade is a magnet for gathering and their umbrage interferes with surveillance. And more recently, in the throes of a homeless crisis, city staff and law enforcement have removed and denuded trees to control the conditions on the street. As temperatures rise, these social and cultural forces that conspire against shade make it difficult for Stephens-Browder and other Angelenos to cool down.

Los Angeles has a long-standing reputation as a warm-weather paradise. Since the nineteenth century, new arrivals have flocked there to chase the sun. The light was believed to be the cure for tuberculosis, a miraculous ingredient to ensure healthful longevity, and later, what turned celluloid reels into movie magic. Sunshine is still the icon of the California lifestyle, drawing us to the beach or the pool to tan, and to cruise Sunset Boulevard with the top down, past the skinny silhouettes of ubiquitous palms. The distaste for shade may originate in solar worship, but it manifests in cultural obsessions like Hollywood noir, in which long shadows and unlit corners represent the criminal underworld.

Shade is not part of L.A.'s modern identity. In the 1930s, the city was rezoned to Federal Housing Administration design standards and banned high-density developments like row houses. Although apartments were once common, city leaders bowed to a prevailing wisdom that L.A. should not resemble a dark and cramped East Coast city. Freestanding single-family homes that were touched by sun on every side became mandatory. In came the cars. L.A.'s curbside trees were removed to accommodate shrinking sidewalks and expanding roads, and new rules that required parking minimums dealt another blow to the urban forest. Mediterranean-style courtyards became endangered species as the shaded commons were converted to outdoor car storage. For decades, no building could be taller than the twenty-seven-story city hall, and even after business elites successfully convinced the city to raise height limits, towers were only allowed in small pockets. "And already their shadows fall in checkered and criss-crossed patterns across the speckled sun-kissed streets of our city," city officials noted in 1959, somewhat ruefully. "The face of Los Angeles is changed and it is lifted with the cosmetics of steel and mortar."

Since the 1970s, an individual right to sunshine has been practically enshrined in state law. Many construction projects fall under the California Environmental Quality Act, a regulation designed for public works that morphed into a tool to thwart unwanted development. Until 2019, L.A. planners required a shadow analysis of any building that loomed five stories over the surrounding landscape, fearing the hundred-foot shadows that could darken nearby gardens and sunbathing decks. Even today, amidst a housing crisis that has pushed tens of thousands of Angelenos like Stephens-Browder onto the streets, city councillors in the San Fernando Valley try to veto new apartment complexes in part because of their shadows. NIMBYs everywhere are quick to complain when their views are blocked, but in California, environmentalists have gone further with the Solar Rights and Solar Shade Control Acts, which protect homeowners from shadows falling on their solar panels. The law even goes so far as to define circumstances in which they can trim their neighbors' trees. In L.A., people have a legal right to sunshine; as for shade, they are on their own.

Given Angelenos' sun obsession, it may come as a surprise that they were once appreciative of shadows. In the eighteenth century, Spanish colonists discovered a cool and lush landscape managed by the native Tongva people in the village that became modern downtown. Such villages were located near the fertile banks of the Los Angeles River, some under great forests of oak trees, sycamores, and willows. After taking the land, the Spanish cleared these woods for arcaded missions and whitewashed adobes oriented away from the sun to keep their occupants cool. The Anglo settlers who arrived in the nineteenth century planted new trees to offset those the Spanish cut down. To compensate for the dry climate, California farmers built a world-renowned system of canals to cultivate citrus orchards. Then the construction of the Los Angeles Aqueduct brought millions of acre-feet of water into the region, triggering a profound ecological change. Naturally bare hillsides and grasslands became dense urban woodlands, and gardens and fruit trees bloomed in backyards across the city. All that moisture cooled the air, and over a fifty-year period, L.A.'s summer highs dropped about a half degree every decade.

The bungalow—a single-story, shoebox-shaped house with overhanging eaves inspired by the architecture of Indian hill stations—became popular with a middle class trying to stay cool. "They were actually prefabricated in factories," said the late Mike Davis, the celebrated scholar of Los Angeles. Tens of thousands of bungalows, many along the Alameda Corridor that runs through South L.A., were manufactured by Pacific Ready-Cut Homes, which advertised itself as the Henry Ford of home construction. Davis, who was briefly a Pasadena urban design commissioner, explained that shade used to be integrally incorporated into the Southern California streetscape. The farm town where he grew up once had its own porticoes, in the form of large storefront awnings hoisted over the sidewalks. Arguably, until the 1930s, Angelenos' relationship to the sun was equally one of avoidance.

All that changed with the arrival of cheap electricity. In 1936, the Los Angeles Bureau of Power and Light completed a 266-mile high-voltage transmission line from the Hoover Dam, supplying 70 percent of the city's power at low cost. Southern Californians bought mass-produced ranch houses with air-conditioning. Shade was outmoded. By the end of World War II there were nearly four million Angelenos in the county, many housed in new neighborhoods organized around driveways and parking lots. Parts of the city became "virtually treeless deserts," Davis said. And as the old farms and orchards were paved under buildings and roads, the air began to warm. Between 1940 and 1990, average and maximum temperatures in downtown L.A. rose about a degree every decade, outpacing other California cities.

Look at what happened to Pershing Square, once a glamorous five-acre park in the heart of the city. In the defining 1910 design, brick-lined paths carved through a dense urban wood. Under the tropical leaves of banana trees and birds-of-paradise, a white-collar lunchtime crowd read newspapers and books from a library cart and congregated to hold forth on global affairs. Then in 1951, the park was bulldozed to install a three-story underground parking garage. The trees were relocated to Disneyland, where the ficuses shaded Main Street, U.S.A., and the date palms became scenery for the Jungle Cruise. On top of the parking garage, Pershing Square became a large and thin expanse of grass, as the subsurface made it impossible to plant deep-rooted trees. To make things worse, the square was fenced off from the promenaders and flâneurs. The park's "nuts" and "blabbers," in the words of the *Los Angeles Times,* were relegated to its edges, where they competed for space with cars entering the garage through deep surface gashes.

It's easy to see how this hostile design reflected the values of the peak automobile era, but there is more than meets the eye. The destruction of urban refuge was part of a long-term strategy to discourage gay cruising, drug use, and other shady activities downtown. In 1964, business owners sponsored another redesign that was intended to finally clear out the "rude panhandlers, deviates and criminals." The city removed the perimeter benches and culled even more palms and shade trees, so that office workers and shoppers could move through the park without being "accosted by derelicts and 'bums.'" Sunlight was weaponized against these undesirables. "Before long, pedestrians will be walking through, instead of avoiding, Pershing Square," the *Times* declared. "And that is why parks are built." But it turned out to be an indiscriminate attack. Shorn of its canopy and surrounded by car traffic, the "see-through" Pershing Square was forsaken not only by scofflaws and degenerates but by shoppers and businessmen, too. It was not so much a dangerous place as it was ignoble, "a sort of last resort for people sleeping off the night before or dozing off the rest of their lives." As *Times* columnist Art Seidenbaum moaned, "Out went sweet shade. In came sterility." The park was abandoned and fell into disrepair.

Yet to a certain kind of politician, the failure of Pershing Square looked like a success. Pershing Square set a template for Los Angeles, whereby a park is not mainly a place to gather and cool down but a revenue-generating asset. Five blocks north of the square sits Grand Park, a rectangular, twelve-acre lawn opened to cater to the downtown's growing residential population. At the opening in 2012, five thousand visitors stood in a freshly landscaped field to watch aerial dancers cascade down the face of city hall, the de facto stage across the street. Movie screenings and concerts are held regularly, and the park rakes in about $1.5 million annually, mostly in rental fees. Like Pershing Square, Grand Park offers scant respite from the sun. It's also built above a parking garage, which leaves no room for underground tree roots, nor for structural footings for a permanent shade structure. "They like the events, they like the lawn, they like being there," said landscape architect Mia Lehrer. "But they find it very hot." In 2016, her firm submitted a winning design for a new park across the street, First and Broadway Civic Center Park, that was informed by critiques of Grand. In public workshops, the city officials who selected her design "heard it loud and clear from people that they wanted shade," Lehrer recalled. The design by Studio-MLA and OMA calls for twenty-six-foot-tall metal structures that resemble California poppies, shading a split-level amphitheater and outdoor restaurant. It remains unbuilt.

Meanwhile, a new renovation of Pershing Square broke ground in 2023. Like previous efforts, it kicked off with a design competition sponsored by the square's surrounding property owners. The winning proposal, by a French firm called Agence Ter, centers on a great lawn with plenty of space for public events. The design also calls for a massive, block-length arbor that the firm calls a "shade pergola." The thin slats of this public canopy will be hoisted thirty feet in the air by clusters of slender columns that expand like tree branches as they extend upward. The whole structure will be scaled by climbing vines, echoing a more natural tree grove that will be planted on the other side of the park. In renderings, the pergola seems to recede into foliage, and crowds mingle underneath in raking light.

But conspicuously, there are no benches. I asked landscape designer Lauren Hamer, then with Agence Ter, about that. Shade creates shelter, she said. "And Los Angeles obviously has a very conflicted position towards creating shelter in the public realm," which is reflected in laws that antagonize the homeless. "Public spaces need to be open, so that people can move across them, as opposed to gathering there." She cited a failed 1986 proposal by James Wines which would have transformed the park into a miniature of the city itself—

a "magic carpet" of different microclimates, each module locked in a grid. Although Wines won an open design competition, his park was never funded or built. Hamer thinks that's because his version of the park was too inviting.

This expectation that Pershing Square not be too social—that it shouldn't be a good place to hang out, as Hamer put it—was strange to the team at Agence Ter. "It's exactly opposite the design traditions that we, as a French firm, know and are used to working with," said architect Annelies De Nijs. "For us, parks are mainly places that are a destination," she continued, where the public comes for a visit and then lingers. The design of the pergola reflects the tension in Los Angeles. De Nijs said the firm mitigated the contested element of public comfort by designing a giant canopy that was "very high and very wide, so it becomes like one big overall ceiling," rather than several smaller and more intimate "shade elements" that unwanted people, like the homeless, might seek to inhabit. Eventually, when the pergola is completed, it will shade a section of Pershing Square in a dimmer and dreamier light. But it will not offer people like Debbie Stephens-Browder relief from the heat.

L.A.'s Shady System for Green Infrastructure

High-concept public architecture is one way to transform the shadescape of Los Angeles. The green infrastructure of street trees, which already does so much for us, is another. Street trees are the unsung heroes of the urban environment. They absorb rainwater to lessen the load on our sewers. They cool the air to reduce the need for air-conditioning and the burden on the energy grid. They clean the air and scrub out pollutants. And a beautiful row increases property values and thus municipal tax revenues.

Because these trees provide so many ecological services, you might think that a warming city like Los Angeles would aggressively support their provision. But for more than a century, city leaders have underinvested in green infrastructure to prioritize the gray kind. There are scant efforts made to protect trees from environmental stressors and unchecked development. Onerous and outdated regulations designed for cars and other urban utilities make it difficult to expand the meager canopy in neighborhoods that need it most. At one-quarter of 1 percent of L.A.'s municipal budget, the funding for trees is a drop in the bucket of the city coffers.

The locals have long desired more umbrageous landscapes. In the 1920s, the chamber of commerce hired the Olmsted Brothers landscape design firm—run by the sons of Central Park's architect Frederick Law Olmsted—to pitch a plan for a world-class parks system. At the time, less than 1 percent of the region was parked, and the green space was concentrated in suburbs like Pasadena and the exclusive communities of Palos Verdes. The firm urged public officials to rectify the situation by purchasing land along the banks of the Los Angeles River and the naturally fertile riparian corridors. Small and large open spaces would be the pearls on a citywide necklace of greenbelts and walking trails shaded by native oaks and sycamores, along with Australian eucalyptuses and tropical palms. Although every Angeleno would have benefited from access to this connected park network, the designers made special mention of the need to serve the poorer residents who were crammed in apartments that had no backyards or gardens.

But the business elites rejected the proposal, denouncing it as government overreach. Los Angeles was chartered with a debt limit and did not so easily issue bonds and raise taxes for sidewalks and sewers, much less a massive acreage of park space. Instead, real estate boosters and homeowners' associations developed their own neighborhoods through special assessments, which are fees they pay the city to make public improvements in the immediate vicinity. This arrangement worked out for the wealthy, but in L.A.'s working-class neighborhoods, "you're just getting a piece of subdivided land, and you gotta do your own thing," explained Marques Vestal, an urban planning professor at UCLA. When federal funding for public infrastructure eventually came to Los Angeles in 1938, city leaders did not use the money to restore the river into a flood-irrigated green channel. Instead, the U.S. Army Corps of Engineers paved over the alluvial plain to protect the important transportation network of nearby railroad tracks, and ever since, L.A. has flushed its stormwater out to sea.

Eighty-three years later, I went to one of those working-class neighborhoods to scout those pieces of subdivided land to find a place to plant a tree. Stephens-Browder's group TreePeople is hired to green Watts, a hard and gray neighborhood deep in South Los Angeles. Although L.A. as a whole is warming, the situation is worse in neighborhoods where there are more dark roads and surfaces, and fewer trees around to shade them. "If they say on the news that it is eighty-five, it is ninety-five here," Stephens-Browder told me, referring to the weather report. "The surface of the buildings, the concrete, cement—all of that absorbs the sun in, almost like boils it, and spits it back out. And that's what we feel." Climate scientists have another name for the phenomenon that Stephens-Browder described: the urban heat island effect.

Some environmental historians say that heat is caused by racist decisions in the past. In 1939, the Home Owners' Loan Corporation, a government-sponsored entity that backed mortgages, created maps of American cities and rated the financial riskiness of different neighborhoods on the basis of their racial composition. In L.A., homogeneously white Westside enclaves like Brentwood, Pacific Palisades, and Palos Verdes were greenlined, which meant they were recommended for investment. But diverse neighborhoods like Watts, which had a large population of Blacks and Mexican and Japanese immigrants, were colored red to dissuade lenders. Redlining denied these residents access to not only mortgages but also loans that could have been used to invest in trees and other environmental amenities that cool the air. To this day, formerly redlined neighborhoods are barer than their greenlined counterparts, and as Stephens-Browder explains, that also makes them hotter.

Because there are few public parks, and the residents don't have big yards, new trees in Watts are often sidelined into sidewalks. Stephens-Browder's colleague took me to a block where he intended to plant bauhinias, umbrella-shaped orchid trees that bloom with fragrant pink flowers. Many sidewalks in L.A. have been repeatedly diminished by various agencies in the name of civic improvement—most notably by road widening to ease traffic, but also by the imposition of urban infrastructure that requires ample clearance. To prevent car accidents, L.A. discourages trees within eight feet of driveways and forty-five feet of intersections. These modern precepts make it all but impossible to re-create the cool and shady streets of yesteryear. (They are also contradicted by evidence showing that streets are safer with trees.) The water meters lodged in metal vaults under the sidewalk require clearance radiuses of six feet, so tree roots don't tangle with the pipes. Street lamps and power poles also come with twenty-foot clearances, so the branches don't obscure the lights or mess with the transmission systems. Then there are the overhead power lines that need twelve feet of room for fire safety. All these things go in the ground before trees, thwarting efforts to expand L.A.'s urban canopy. On this block of forty-eight houses, fewer than a dozen were eligible for a new curbside tree.

Then comes the harder work of growing those trees to maturity. To survive the harsh conditions of an urban sidewalk, where soils are compacted and roots don't have much room to grow, trees need a lot of care—three to five years of consistent watering, regular pruning, plus staking, mulching, and weeding. Yet Los Angeles does not tend to its trees as attentively as it does to other urban infrastructure, like roads. Although city crews take twenty-one years to trim a tree, they need only three or four days to fill a pothole.

The arboreal abdication does not bother Scott Goldstein, a Hollywood screenwriter and director who stewards the street trees in Windsor Square, a historic district in the middle of the city. His luscious urban forest is visible on satellite maps as a green island in a sea of gray streets. This is a neighborhood that was planned for shade. The homes were constructed on spacious lots, easily two or three times the size of those in Watts, that residents filled in with verdure. The sidewalks were uniformly lined with broad planting strips and the telephone wires were undergrounded to allow the crowns to flourish. On a warm Sunday morning, as lawnmowers murmured, Goldstein took me on a walking tour. The streets tunneled under rubbery southern magnolias and the handsome canopies of California oaks. We stopped to admire the jagged leaves and flowering red berries of a toyon tree. Naturally, these shrubs blanket the slopes of the Hollywood hills. But even natives need nurturing when they are crammed into concrete. In Windsor Square, a sapling spends its first two years gradually watered by a drip irrigation tube. It receives regular soaks from an overhead sprinkler. Then, at maturity, when the roots threaten to lift the sidewalk, residents shell out a few thousand dollars to mitigate the conflict.

Many American cities ask neighbors to water street trees, but the request is imperative in L.A. because of low rainfall and long droughts. Goldstein did not fault the city for failing to lavish its trees with care. "You have a huge city like Los Angeles with a huge urban canopy. How do you possibly take care of that?" he asked me. "The homeowners can do it," he went on, "but they have to want to do it and they're going to have to spend the money." The money is not an issue to the residents of Windsor Square, who hire landscapers. But most Angelenos do not have that dough.

Down in Watts, Stephens-Browder has to convince the neighbors of the worthiness of the investment. TreePeople, like other greening nonprofits, is not typically funded for maintenance, and asks volunteers to do the work the city cannot. For this reason, most residents pass on new street trees. They don't want the personal responsibility for this public amenity, and it takes only a stroll down a crumbling sidewalk to see why. When trees aren't pruned, their branches tangle with power lines, and their overgrown crowns engulf lampposts, shrouding the streets in nighttime darkness. Worst of all, the roots can tear into plumbing, or thrust out of the ground, destroying the pavement. Surprisingly, those costs are borne by the adjacent property owner, not the city. Volunteerism has many benefits, but when it comes to urban trees, it's a lousy policy, and even worse, it entrenches inequalities.

Stephens-Browder is not fazed. She tries to win over Watts residents by reminding them about a different problem. Angelenos everywhere are stressed by heat waves, but after five days of extreme temperatures, "that's when people start to keel over, and it's people in Black and brown communities that are affected most," she said. "This heat is serious." During a heat wave, Watts residents are twice as likely as the average Angeleno to end up in the hospital and six times likelier than someone in a Westside enclave. The neighborhood's elderly Black and Latino residents are more likely to perish. The unequal impact is an example of what University of Southern California researchers call the "climate gap," or the tendency of climate change to burden the poor and people of color. Watts residents suffer not only because of the higher temperatures, but also because they cannot afford the air-conditioning needed to cool down. The heat also worsens health conditions that are more prevalent in Black and Latino communities, such as asthma, diabetes, and heart disease.

Scientists believe trees could go a long way to cool Watts and even close L.A.'s climate gap. But until green infrastructure like street trees is afforded the same room on Watts sidewalks as gray infrastructure like driveways and storm drains, the efforts of Stephens-Browder and other foresters may not amount to much. To achieve the aspirational shade that could save lives in Watts, trees cannot be a lightly supported and nonprofit-driven passion project, but a form of mandatory urban infrastructure that is nearly as important as the almighty car. Trees must be elevated in the streetscape with new rules and regulations that command their placement along roads. Every time a street comes up for repair it must be standard procedure to spare new space for a tree, especially in neighborhoods that don't already have them. Trees should be one of the first things to go in the ground instead of the last.

And if the city's politicians are serious about trees as the public's first line of defense against rising temperatures, then they will have to invest in trees as green infrastructure on the order of gray infrastructure like sewers and highways—as bond-funded projects with large budgets, up-front capital costs, and ongoing maintenance. City leaders cannot accomplish this with an annual provision of $25 million, which is what L.A. spends on trees today. According to the L.A. Urban Forest Equity Collective, a research project funded through a public-private partnership, a greener, shadier, and fairer L.A. would begin with a long-term investment of at least $664 million. That might sound like a lot of money, but averaged over a decade it would not be out of proportion with what L.A. budgets for gray infrastructure, and far less than the $175 million spent every year to maintain the road network and the $370 million to expand the sewers.

This is not to say that trees are more important than roads and sewers. But if Angelenos could at least think about trees as infrastructure on a similar tier, then their costs could be more acceptable. For now, L.A. is content to allow individual residents to make their own shade with their own nickels and dimes.

When It Comes to Shade, City Leaders Say Their Hands Are Tied

Not long after canvassing Watts, I took the subway through a mountain pass and stepped out in the San Fernando Valley, a land of spacious lawns, tidy ranch houses, and prostrating heat. Blocked from the Pacific Ocean's cool breezes and cloudy marine layer, the Valley bakes under the California sun. It was two weeks into fall, and the forecast called for an afternoon high of 96 degrees. Nevertheless, L.A.'s then-Mayor Eric Garcetti was in a chipper mood when we met on a North Hollywood street, after a press conference to announce a new initiative to combat the urban heat island effect with solar-reflective road surfaces. "Most people don't like to throw shade," he told me, grinning from ear to ear. "I love it."

Garcetti achieved international renown for co-founding the Climate Mayors, a consortium that committed to the emissions goals of the U.N.'s Paris Agreement in their cities after the formal withdrawal of the United States. More than any American mayor, he made shade a plank of his agenda. Garcetti harbored an ambition to reduce L.A.'s infamous car dependency and believed comfortable sidewalks might entice more Angelenos to choose a lower-carbon lifestyle. On the advice of his chief design officer, a former *Los Angeles Times* architecture critic named Christopher Hawthorne, Garcetti began talking about shade as a new kind of infrastructure, in forms both green and gray, that could also make the streets more bearable to those who had no choice.

"Maybe you haven't thought about it this way, but shade is an equity issue," he declared two years earlier. "Think about an elderly Angeleno who relies on public transit to get around her neighborhood," he went on. "Imagine her standing in the blistering sun in the middle of July waiting for the bus, with hot, dark asphalt. She deserves to be every bit as comfortable as her counterpart in another zip code in town," who could endure the heat in an air-conditioned car. In the era of climate change, shade could not be a luxury, but more like a civil right. By the end of 2020, Garcetti planned to shade 750 forlorn bus stops with trees, umbrellas, sails, or architect-designed structures that slid over benches. He also promised to plant ninety thousand new trees across the city by 2022 and expand the tree canopy by 50 percent in the hottest and poorest neighborhoods by 2028—the foundation of a groundbreaking plan to achieve shade equity in Los Angeles.

But granting the right to shade was easier said than done. When I met Garcetti, he was nearing the end of his second term, and those goals were not in sight. The architects' bus shades were unfunded and on indefinite hold. The trees were slow to come in. Even if every one of L.A.'s empty tree wells were filled and front yards were foliated, it still would not achieve the mayor's definition of shade equity. The mayor's ambition required the extensive reconstruction of a huge number of city streets to strip parking and traffic lanes for trees—a literal breaking of ground Garcetti's administration neither planned nor budgeted for. Garcetti had another "issue" besides shade: a history of overpromising and underdelivering. In his first term, he pushed a major overhaul of L.A.'s transportation plan through the city council to require wider sidewalks, new bike paths, and fewer traffic lanes all over the city. The moves could have laid a literal foundation for more shade, but after loud objections from drivers and firefighters who claimed narrower roads would worsen traffic, the council turned its back on Garcetti's plan. Only a fraction of city streets were actually rebuilt.

What happened? "People think of L.A. as liberal. We're libertarian," Garcetti told me. The public works philosophy was to "do the basics, build the freeways, bring some power, and get out of the way." He reminded me of the assessments that homeowners and developers used to petition the city for infrastructure in their neighborhoods. To this day, he claimed, L.A. is the only city in America with assessments for even the basic service of electricity. "If you want streetlights, and if you want to keep the power on, you literally have to keep the neighborhood voting to pay for their power themselves," he said. In other cities, taxes fund these services for everyone, but in L.A., Garcetti said residents ordered them, like items on a menu. As we talked, he mused about his mayoral successor tapping a $120 billion county transportation fund to build shadescapes for decades to come. Evidently, there was money to change L.A.'s streets for the better, but he seemed to be saying that constituents weren't asking their elected officials to do it.

Felipe Escobar, a community organizer in the San Fernando Valley, offered a different view of the political situation. Escobar's group Pacoima Beautiful works with families to demand the basic infrastructure that the neighborhood's earliest industrial developers were never assessed for, like storm drains, paved sidewalks, and trees. Pacoima isn't just one of the hottest neighborhoods in the city: It's also one where residents spend more time outside, which prolongs the exposure. Escobar took me to a dusty intersection near the train tracks where elderly women waited for the bus. To protect themselves from the sun, they hid under striped umbrellas and in the slivers of shadows cast by utility poles. Or the riders wander away in search of shade, and when they're gone, the bus drivers skip the stop. A thirty-minute wait can become an agonizing hour. Pacoima residents are seven times likelier to end up in the ER on a hot day than someone in the cool Pacific Palisades, and the disparity will widen as the San Fernando Valley warms. "We don't have air-conditioning, and our parks don't have tree canopies," Escobar explained. Without those defenses, "we're on the front lines of climate change."

Several years ago, Escobar and a handful of neighborhood mothers surveyed riders at bus stops like this, asking what they needed for a safer trip. Everyone said shade. But when Escobar asked transit planners to bring shelters to ten stops, they could only service four. "I've seen where the money goes," Escobar said. "We want to make sure the streets are paved well. We want to make sure the freeways are big enough for everybody." The funds that transportation officials used to widen a few miles of the Sepulveda Pass, a busy Westside freeway, would have been more than enough to install shelters at thousands of bus stops throughout the region—something Angelenos have been promised since the 1980s. "We think everyone drives," Escobar continued, but in neighborhoods like Pacoima, "we have a lot of people who ride the bus." Almost all of them are Latinos living in poverty. They can't afford lobbyists, and they can't take off work to speak at city council meetings. And they aren't afforded resources to adapt to heat.

Rather than use the city's significant transportation funds to meaningfully shade the streets of the Valley, Garcetti instead acquiesced to the status quo of fobbing those costs onto residents who could afford them. Perhaps that makes sense from a municipal finance perspective, but when it comes to tackling shared concerns like climate change, it is not an effective strategy. Moments before we met, I watched Garcetti speak in fluent Spanish to an elderly woman named Olga Hernandez, who watched the press conference from her doorstep. She didn't care much about the reflective street paving. She wanted a sidewalk and told Garcetti she had been asking for years. Isn't that what her taxes were for? What wasn't she understanding? A student of L.A. history, Garcetti explained to Hernandez that when her neighborhood was built, the ranch style was in, and some developers didn't bother with sidewalks. Now, he continued, it would be too expensive for the city to foot the bill. Garcetti turned to her councilman, Paul Krekorian, who was standing nearby, and asked what it would take to set up an assessment to corral those costs from Hernandez and her neighbors. It was fascinating to watch this famous climate champion pass the buck in real time explaining why he could not provide for his constituents one of the most basic services imaginable to cut carbon emissions: the opportunity to walk rather than drive. But then, there was a lot that L.A. could not do, like find shelter or shade for Debbie Stephens-Browder.

Shade as a Matter of Public Safety

Stephens-Browder remembers a childhood outdoors. She and her cousins spent the summers in verdant backyards, picking loquats and lemons from family gardens. Magnolia trees lined the sidewalks of her block. "It was lush," she recalled, her eyes lighting up behind thick glasses. "Every time I dropped something, a tree grew." Parts of Los Angeles are located on a migratory path of monarch butterflies, and Stephens-Browder believes her childhood home in Compton and her grandparents' old house in Watts are on that route. One particularly sweet tree seemed to be filled with the winged insects, so much so that the family called it "the caterpillar tree." "If you have a tree called a caterpillar tree, can you imagine how many butterflies there must be in that community?" she asked me. But these days, she never sees a butterfly in Compton or Watts.

Stephens-Browder believes the butterflies left South Los Angeles because they lost their habitat of trees, and she believes the trees may have died because the police sprayed them with pesticides. Although this theory is unproven, it has some currency in Watts. I first heard it from Tim Watkins, the president of the Watts Labor Community Action Committee (WLCAC), a nonprofit community association in whose parking lot Stephens-Browder sometimes slept. After the Watts riots of 1965, Tim's father, Ted Watkins, used federal money and a loan to buy dilapidated lots and redevelop them as housing and parks. One lot became a tree nursery, and in the 1960s and 1970s, the organization hired neighborhood kids to plant those trees under a city contract. Many of them are still standing, like the "monstrous" windbreaks Tim remembers planting on Pacific Coast Highway and thick stands of ficuses in Hollywood and the San Fernando Valley.

Some trees went on Central Avenue in Watts, because if trees were "good for other communities, then Watts needed to have its share, too," Watkins said. But most of them are now gone. Years ago, he took me on a driving tour and pointed out vacant tree wells in front of a senior center operated by his organization. I suggested that traffic engineers prevented city crews from planting new ash trees, but Watkins offered a more startling explanation. "We're told that, because of the crime rate in Watts, and the need for police helicopters to see who was under the trees, they started a defoliation program," he later told a radio host. "Even today, with all the efforts over the years to restore that, we still have whole blocks where trees were once planted where there are no trees now." Watkins links the defoliation program to malathion, a chemical that state helicopters sprayed on L.A. trees in the 1980s and 1990s to rid them of the Mediterranean fruit fly. Watkins believes the campaign may have served an additional purpose of denuding the urban trees whose shade sheltered criminals.

Requests to deforest are common in L.A.'s high-crime areas, where shade is perceived as concealment for drug deals and turf wars and a magnet for modern-day *umbratici* like drunks and prostitutes. Police captains, neighborhood cops, and city attorneys conspire to remove criminal cover and urge the city's overburdened streets department to whack away canopies that harbor these allegedly dangerous elements. Often, it's their constituents pleading for help, as when tree branches and leaves block out streetlights. But there's something about daytime shade that seems to draw law enforcement's ire. As one South L.A. officer put it, criminals are "creatures of habit." Get rid of their shade and they will no longer congregate. The approach on private property is less formal. "It's not that they have the authority to say, you can't plant trees here," explained Michael Pinto, an architect who helped design an urban farm for Watkins. "It's that they have convinced community leaders that, if you want to save your community, you can't have too many trees, because it restricts their ability to do their jobs."

The belief that trees are accessories to crime is as old as the hills. In the thirteenth century, a British royal edict commanded two hundred feet of forest clearance on main roads as a precaution against hidden robbers and burglars. Susan Phillips, a Pitzer College anthropologist who studies L.A. gangs, reminded me that American armed forces more recently deforested enemy territories. On the Western frontier, Kit Carson burned the Canyon de Chelly to force the Navajo into surrender, and the U.S. Air Force torched the jungle to deprive the Vietcong of cover for ambushes. These efforts may even have inspired Saddam Hussein to dry the marshes of southeast Iraq, thought to be the biblical Garden of Eden where Adam and Eve, after their original sin, hid from God in the verdure.

But the tradition of killing trees in the name of public safety is somewhat newer in Los Angeles. Rather than invest in the basic services of smooth sidewalks and shady trees, city leaders instead fund police to the tune of $3 billion a year and charge them with the upkeep and enforcement of the public realm. Since 1995, the Los Angeles Police Department (LAPD) has been steeped in the principles of Crime Prevention Through Environmental Design (CPTED), which approaches public safety through the lenses of architecture and urban planning. Many American law enforcement agencies have CPTED task forces and circulars, but L.A. has gone a step further by assigning police officers to building plan reviews where they suggest changes to the lighting and landscaping. Officers also use CPTED principles to guide their safety assessments of public housing communities and crime-ridden parks. Because gangs stash their drugs in tree canopies to avoid detection and hide from cops behind bushy shrubs, the police cut them down in the notorious Imperial Courts, said Perry Crouch, a longtime Watts resident and peace activist. Sometimes, the real concern isn't concealment, but the comfort shade affords unwanted presences. Infamously, during a broiling 2023 summer, Universal Studios hat-racked a row of ficuses that shaded a picket line of striking screenwriters and actors. The studio dubiously claimed the pruning was prescheduled.

The core principles of CPTED emerge from *Defensible Space: Crime Prevention Through Urban Design,* a 1972 book by architect Oscar Newman that offered basic design tenets to improve public safety. First, establish territoriality, which is the informal ownership of public space. Second, encourage natural surveillance and make it easier to see criminals. Third, remove graffiti, litter, broken windows, and other visible signs of lawlessness. CPTED practitioners also impute a fourth principle of "access control" from Newman's work. Those ideas don't seem to have much to do with shade, but in the decades since they have come to impart a "negative message about trees," as the writer Richard Conniff observes. For instance, a billowing camphor that shades a window could also block a homeowner's view. A burglar could lie in wait behind the trunk. And overgrown branches could be a visible sign of disorder and encourage more crime. If trees must be planted, CPTED practitioners want the branches trimmed and the lower limbs removed, but in L.A., a former city official told me, the neighborhood police would prefer not to have them at all.

"Natural surveillance is easy," explained James Nichols, a veteran street cop who runs the LAPD's CPTED unit. "That just means you have to have the landscape pretty much clear." An amiable fellow, Nichols hosted me in the police department's downtown headquarters. His office overlooked a neat and grassy plaza that stepped down to the street. "Criminals want to exist in the shadows," he continued. "The less chance they get to be identified, the happier they are." After showing me the ground floor blueprint of an apartment building where he recommended more lights and cameras, he began to thumb through illustrative photos in a CPTED training manual. "He's standing right out in the open, but you can't see him because he's in the shade," he said, showing me an assailant standing in the shadow of a coniferous tree. "It takes time for your eyes to adjust, and then you'll see him."

"And then it's too late," corrected his partner, Officer Alfonso Velasco. It seemed to escape Nichols and Velasco that in L.A.'s consistently warm climate, shade might support the department's goal of public safety rather than detract from it. There really is something to the expression "hot under the collar." Heat is an irritant that makes us angrier, more aggressive, and even more vengeful. It can override more rational thoughts. Craig Anderson, a psychologist at Iowa State University, has observed that high temperatures induce a "cognitive priming effect, automatically increasing the accessibility of aggressive thoughts." We interpret ambiguous social interactions more negatively. We think the worst of others and give in to our worst impulses. A minor altercation can escalate quickly.

Economists, criminologists, psychologists, and social scientists all believe heat can cause more crime. Reams of associational studies confirm that antisocial behavior rises with the mercury. Mail carriers say they're harassed more when the temperature is above 90 degrees. Drivers honk longer and are more liberal with the bird. Baseball pitchers intentionally hit more batters. The association with gun violence may be strongest of all. Although assaults are more common in summer because there are more interactions outside, even a balmy fall day can turn deadly. In Los Angeles, university researchers and economists have urged city leaders to trial "novel solutions" to reduce shootings, like nighttime events and temporary gang cease fires. Kelly Turner, a UCLA urban planning and geography professor, suggests more shade. Denuding a row of trees might deter a criminal from breaking into an apartment by robbing his concealment. But the same act also makes the whole block a little more dangerous by making everyone else a little angrier. If L.A. cops stepped out of their air-conditioned cruisers and spent more time on foot in the sun, they might think a little differently about policing shade.

Since the early 2000s, police captains, gang detectives, and neighborhood cops have relied on security cameras to monitor situations in public housing, parks, and other sensitive locations. In 2005, the LAPD began installing Motorola video cameras on power poles in high-crime areas and asked city crews to cut back trees that obscured their views and interfered with transmission. Eventually, street cops submitted so many requests that the forestry department started recommending tree removals in places where regular maintenance was not feasible. Today, the courtyards of Watts's housing projects are shorn. Additionally, when South L.A.'s Harvard Park was redesigned to become more "family friendly," as a veteran gang detective described it, the mature canopy around pole cameras disappeared to ensure a "clear visual." The detective didn't believe the cameras deterred crime, but when it came to catching the criminals, he added with a chuckle, they "worked like a charm."

Aaron Thomas, a veteran urban forester, said the webs of sight lines required by surveillance cameras make it difficult to find places to plant new trees. Thomas is hired by the Housing Authority of the City of Los Angeles (HACLA) to spruce up the buildings, but the impractical guidelines result in barren landscapes with fence-like edges. "Either you're going to prioritize making a community more resilient, or you're going to prioritize surveilling them," Thomas said. And that's a shame, because in Watts's public housing communities, the townhouses open onto large, spacious lawns that could easily support towering sycamores and live oaks. But it doesn't happen, said Cynthia Gonzalez, a former HACLA project manager, because "law enforcement associates safety with cameras." During her tenure, she said, her colleagues disputed claims that the ecological benefits of trees outweighed the risks of obstructed views.

Gonzalez understood the need for protection and didn't minimize the fact that Watts's public housing residents deserve to feel safe. Nevertheless, other cities have figured out how to accommodate shade trees in public housing without sacrificing security. A study of New York City's urban forest found that housing communities in the Bronx, Harlem, and Brooklyn had far more tree cover than the surrounding neighborhoods, due to expansive lawns and courtyards that were planted decades earlier with London plane trees and elms. In fact, the outdoor temperatures in the projects were 2 degrees lower than in the surrounding areas. From a health perspective, during a heat wave, these courtyards may be some of the safest places in the neighborhood.

In L.A., the conflicts between shade and security could be resolved if the city funded its forestry department for routine pruning. Rather than allow trees to grow out of control, obstruct security cameras, and become targets for removal, the city could trim them every five years as professionals recommend. In Chicago, a much-cited 2001 study of the Robert Taylor Homes found that public housing with trees had 52 percent fewer crimes and 56 percent fewer violent crimes than buildings isolated in plots of pavement. The trees were carefully pruned to avoid the overgrowth that could conceal a criminal. Additionally, university researchers in Baltimore and Portland found shadier streets had lower crime rates than sunnier ones, all other things being equal. It was just an association, but a strong one. It may seem intuitive that when more people are outside on hot days, they are more likely to get in someone's face and cause a confrontation. But according to the researchers, the shade brought more people outside to hang around, and their presence acted as a deterrent.

Nevertheless, the notion that shade does not create crime but prevents it remains a tough sell in Watts. Residents are desperate for help, and if that help comes in the form of more lights and cameras and fewer trees, so be it. On my driving tour with Tim Watkins, we went to the site of his future urban farm, across the street from one of those public housing projects, Jordan Downs. He showed me a massive tree on the edge of his property that created a shady tunnel over the sidewalk. Police asked him to remove it because "loiterers hang out under the tree and the helicopters can't see them," he said. Eventually, he would oblige. No need for malathion.

"If we say food insecurity, do we say home insecurity?" Stephens-Browder asked me. It had been five months since we first met, and she was no longer so cheery. Car troubles were getting to her. She was losing faith. "I've been to every organization," she said, to find a home inside. I heard weariness and frustration in her voice. We were having lunch at a restaurant two blocks away from Ted Watkins Memorial Park, a neighborhood hub named for the local hero. The park was bordered by mature elms, sycamores, and jacarandas, and sometimes Stephens-Browder parked her car there overnight to wake up in the shade. But because there was no gate, she felt unsafe and vulnerable to attack.

As police in South Los Angeles seek to remove trees to deter loiterers, prostitutes, and gang violence, the police in the San Fernando

Valley are fixated on another perceived threat to public safety: the forty-five thousand people, like Debbie Stephens-Browder, who live on the city streets. Many of them have been priced out of their homes. Others are substance abusers and a significant portion suffer chronic pain and mental illness. Nevertheless, the sight of so many tents across the city does not elicit sympathy so much as fear and anger. Some of it is warranted. The Los Angeles Fire Department responded to nearly fourteen thousand fires at encampments in 2023, which endangers not only the homeless but also the nearby housed residents. The homeowners are bothered by the unsanitary conditions of their streets, threatened by the outbreaks of violence, and aggravated by L.A.'s apparent inability to address a shameful crisis. During the coronavirus pandemic, police urged these residents to denude their neighborhoods, in an apparent belief that sun would force the unhoused to desert the sidewalks and parking lots and take their problems elsewhere.

"Nothing I can do legally about making the homeless go away," wrote a police officer named Jose Moreno in an email to a constituent in August 2020. At the time, the city had suspended sweeps to prevent the spread of the virus and allowed the unhoused to shelter in place. The public health measure frustrated the residents of Topanga and Canoga Park. Desperate homeowners, library staff, and postal workers emailed Moreno for assistance, begging him to trash the encampments. "The only other suggestion I can offer," Moreno wrote, "and it's one that I provide to a lot of businesses who have the same complaints as the library, is to take away the comfort level." He urged neighborhood residents to trim any tree that cast shade or bush that offered visual cover. "Any shelter or hiding spot that can be removed, should," he wrote. "This will at least force them to look for another place."

In nearby North Hollywood, a concerned resident reached out to a neighborhood lead cop named Jonathan Ojeda to see about mowing down the vegetation in the Tujunga Wash, a natural tributary of the Los Angeles River. "Maybe if we clear the view from all the houses, they will leave," the resident wrote, referring to a nearby encampment. In a northwest San Fernando Valley park, an officer named Sandra Zamora confronted a man who gathered his belongings in the shade of a tree: Leave now or face a citation. More officers on patrol swept through to urge the park's other homeless users to relocate ahead of a tree trimming. Not far from Moreno's turf, a resident reached out to a city councilman named John Lee to see about moving overnight RVs off the street. Initially, staffers suggested restrictions to tow the vehicles. Unable to wait, the resident trimmed his trees to remove the shade and asked the officials to install more lights. For his part, Moreno brought his ideas about "the comfort level" to the council office of Bob Blumenfield. There, the policeman found the sympathetic ears of aides and deputies who themselves had previously asked city crews to remove the "shade effect" from trees that attracted encampments to parking lots.

It is surprising that civil servants would undermine L.A.'s stated goal of protecting and expanding the urban forest. Since 1996, a landscape ordinance has required shading at least 50 percent of parking lot surfaces to mitigate urban heat and encourage stormwater retention. Three consecutive mayors have attempted to plant more trees in neighborhoods that need them most. The public voted for the mayors and councillors who passed these policies, which makes it antidemocratic and counterproductive for their own staff to hinder the efforts. But because the homeless are L.A.'s personae non gratae, the outrage is limited. In July 2023, when Elizabeth Chou, a reporter with *Los Angeles Public Press,* published texts and emails showing that city officials and police conspired to remove a tree in a West Hills park as part of a secret plan to break up an encampment, the city shrugged. Blumenfield expressed contrition for the actions, but the neighborhood residents urged the police to finish the job.

Although police watchdogs decry the degradation as an instrument of surveillance, the more immediate concern is that homeless people are already vulnerable to heat, and shade removals increase their exposure. The sun dehydrates them, and even if they could find a tap for water, drinking would eventually force them to pee on the street and risk a citation. Because they stand all day, blood pools in their feet and ankles. They sweat through socks, and blisters and ringworm fester in their feet. The lucky ones make it to clinics where doctors may prescribe antibiotics like tetracycline and doxycycline that increase photosensitivity, causing rashes, itches, and scarring on exposed skin, even as they treat infection. On sultry nights, sleep is elusive. Those who make it to morning could wake on a scorching sidewalk that sears off their flesh. "They seem more frustrated," said Coley King, a doctor and street medic who treats the unhoused. "They don't have a way out." The stress can trigger a psychotic break or a craving for drugs, and a relapse only offers a brief reprieve before things get worse. Alcohol, opioids, and stimulants can raise internal temperatures to hyperthermic levels, and sometimes, they never come down. In Los Angeles, the unhoused account for nearly half of all heat-related mortalities, and in Phoenix, officials estimate that someone who lives on the street is two to three hundred times more likely to bake to death than someone who lives inside.

Given the dire health threat, what could possibly explain the police's sadism? Are they so complacent about homelessness that they have become indifferent to the suffering? L.A.'s unhoused advocates blame the city's hostile attitude toward the homeless. The city has long had rules that prohibit sitting or sleeping on streets, but in recent years the council has amended the municipal code to create thousands of "special enforcement zones." Perhaps by engineering uncomfortable public spaces, the police are merely making it easier to enforce the law. But as Angelenos learned in Pershing Square many decades ago, a denuded park is less inviting for everyone.

"Where does it end?" asked Debra McNelley, one of the residents of the West Hills park encampment. "Where does the discrimination end?" McNelley didn't want to live on the street, but when she could no longer afford her rent, she had no choice. She stuffed her home into cardboard boxes and grocery carts and dragged what remained of her belongings under a tree. She did her best to maintain her home in the shade. And yet it was for this crime of living there, she said, that the police sought to punish her. "They're willing to go to any length to get us out, regardless of how it affects us physically, mentally or psychologically," McNelley alleged. Imagine what could happen if, instead of conspiring to harm the homeless, the city found ways to help them.

When we defoliate our public spaces so the homeless can't benefit from the shade, we don't just spite one undesirable group. We take a resource away from an entire community, and we deny opportunities for people, even those who live on the street, to participate in public life. It is no different from removing a bus bench so no one can sit there, or throwing boulders on sidewalks so no one can hang out there. Arguably, shade removals are more harmful, because of the physical threat of heat. During the civil rights movement, some municipalities chose to close their pools rather than integrate them by admitting Black people. Affluent residents went ahead and built their own pools in their backyards, concentrating summer cooling in the hands of a few. It is not hard to imagine a similar outcome in the twenty-first century with urban trees. Every time we weaponize sunlight, we make shade more of a private luxury and less of a public resource to be shared by all.

Cynthia Gonzalez, the former HACLA project manager, would agree. When we talked, the Pardee RAND Graduate School professor and lifelong Watts resident looked out the window of her childhood home onto a backyard forest of avocado, pomegranate, and citrus trees. Like Debbie Stephens-Browder, she is fortunate to have been raised in a shady oasis. "Communities of color, we get the climate change issue," she said, naming a few ways that her family and neighbors beat the heat. They hang out on the patio. They cool down in the park. Many Watts locals are environmentalists by default. Because they can't afford air-conditioning, they shade their home with trees. They watch their water use to avoid a big bill. Plenty of them recycle and turn those bottles and cans into green in their pockets. "What we can do individually, we've definitely done all that," she added, but individual actions may not be enough to protect entire communities from rising temperatures.

At HACLA, Gonzalez managed Watts Rising, an effort to upgrade Watts's environmental infrastructure as the aging public housing stock was rebuilt. The program was seeded by a $33 million state climate grant that funded not just new housing but also rooftop solar panels, home energy retrofits, electric vehicle charging stations, and four thousand new trees in streets, schoolyards, and parks. With some care and protection, the wispy African fern pines could one day grow into towering canopies that shade the cinder block townhouses. But it's going to take decades for them to come in. What about the people who need shade now?

When her car is in disrepair, Stephens-Browder must take the city on foot. As she winces beneath a piercing summer sun, sweat begins to pool on her brow. Tonight, she hopes to sleep inside, but in the light of day, the enclosure of a bus shelter may be her only reprieve. Not far from Pershing Square, a trapezoidal median is clenched between two major roads and thirteen lanes of traffic. For decades, this island was where the residents of Glassell Park boarded a streetcar that shuttled them to shops and grocers. But when the transit system was disassembled, the parcel was designated for utilities. As the streetcar morphed into the modern-day bus, new passengers came to wait for their rides in the shadows of a telephone pole and transformer boxes.

In 2004, a nearby resident named Helene Schpak decided the riders deserved a more dignified experience. She planned to request a shelter from the local council office, but then she met Michael Pinto, the architect who worked with Tim Watkins in Watts. Pinto's students at the Southern California Institute of Architecture (SCI-Arc) designed a bespoke solution for the site. In the original design, a steel frame structure supported a row of five triangular white sails. Riders were to have unobstructed views of bus paths on two sides of the island. The outdoor pavilion would shade every passenger waiting for the bus—far more than the handful of riders who squeeze into the city's standard models.

To convince the city to accept the shade structure, Schpak pitched a pop-up tent on the island and collected 650 commuters' signatures. Then the council office and the public works department informed her of the costs of restoring the land's original public use. First, it needed new curb cuts to become wheelchair accessible. To accommodate the structure's underground foundation, a gas line had to move and electrical wiring had to be rerouted. The budget ballooned to $190,000, then to $237,500, and then to $630,000, before the final cost was tallied at $352,470. "That's a hell of a lot of money for something that we thought was simple," Schpak told me.

For years, the site was fenced off as crews reconfigured the utilities. During that time, the city endured summer after summer of record highs, breaking a glass thermometer downtown. Finally, thirteen years after Schpak first asked about a shelter, she attended the opening of the revamped Glassell Park Transit Pavilion in 2017. It did not have the lightweight, Erector-set look that Pinto's students once imagined. Instead of five sails angled for maximum effect, there were three—gray, not white. But there was shade. Schpak thanked Pinto, and her city councilman, and the members of the neighborhood improvement association. They saw what needed to be done and never squinted as the city changed around them.

Chapter 6

The Heat Dome: In the Cool Northwest, the Sun Turns Deadly

It came from the other side of the globe. In June 2021, a strong mass of humid air rose from a tropical cyclone, floated across the Pacific Ocean, and got sucked into a jet stream, one of the high-altitude wind currents that circle the planet. Jet streams form between the tropics and the poles, and the difference in temperature creates the shape and velocity. Typically, a jet stream whistles through the atmosphere, pushing weather systems eastward, but the unusual burst slowed the efficient movement. The jet stream began to wobble. It curled around a ridge of atmospheric pressure above the United States and Canada, and held it in place.

The ridge settled over the Pacific Northwest and captured desert winds moving north from Arizona and Nevada and west from the Cascade Mountains. Heat usually rises, but the atmospheric pressure pushed the warm air down to the surface and trapped it there. In normal conditions, moisture in the soil would have cooled it through evaporation, but that spring was the Pacific Northwest's driest since 1924. Instead, the parched soil absorbed the sun's energy and transferred more heat to the air. This all happened a few days after the summer solstice, the time of year when the receipt of solar energy is most direct. One of America's cloudiest regions became one of the brightest, and the foggy city of Portland, Oregon became an urban desert.

When you think of heat, Portland is not a city that comes to mind. It's cool, misty, and mossy, with gray skies nine months a year. It's a nice day when the sun peeks through. Because of the mild climate, the National Weather Service hesitated to issue a forecast calling for temperatures that would be extreme for Phoenix, Arizona. The previous summer, a triple-digit prediction had proved premature after nearby wildfire smoke blotted out the sunlight. Fires were burning again this year. But this time, the aerial shade did not materialize. The atmospheric pressure pushed away the smoke and clouds, allowing the sun to strike at full bore. Meteorologists called the situation a "heat dome."

On Wednesday, June 23, Multnomah County regional health officials sprang into action, urging the public to spend the next few days hunkered down inside with air-conditioning. "This is life-threatening heat," an official stated. About 20 percent of Portland's 630,000 residents did not have AC because they rarely needed it. Some fled their homes for air-conditioned hotels. Others, in desperation, taped cardboard, dark blankets, and aluminum foil to their sun-facing windows. If someone absolutely had to go outside, and saw someone else who looked disoriented, they were advised to call emergency services.

Like other West Coast cities, Portland has thousands of unhoused people who live in their cars or on the street. Officials arranged to open three air-conditioned cooling centers and directed all who didn't fit to seek daytime shelter at churches, libraries, and parks. The largest public refuge, at Portland's convention center, opened on Friday, June 25. Although the official temperature outside was only 95 degrees, EMS crews had to treat a high volume of people in need, including young men and women, in their twenties and thirties, who collapsed at bus stops and passed out on sidewalks. Summer camps, farmers' markets, a cycling fair, the outdoor opera, food carts, and vaccination sites all closed. The parks department revoked its permits for ball games. Oregon is one of two states in the country that require station attendants to pump gas. The attendants vanished after the fire marshal suspended the regulation. The Rose City was shutting down.

Then it began to break down. On Saturday, June 26, Portlanders woke to a piercing blue sky and started dialing 911. People who braved the outdoors or had nowhere to go were found in doorways, alcoves, and behind vending machines, hiding from the sun. That afternoon, the city recorded an all-time high temperature of 108 degrees. The sun began melting a power cable on the light-rail system, which was free that weekend to offer air-conditioned rides. It started to buckle a road in the Kenton neighborhood, leaving a fifty-foot gash in an intersection.

On Sunday, June 27, Portland General Electric set a record for electricity demand, shattering the one set during a frigid December evening in 1998. Demand would have been higher if the system hadn't overloaded. More than 6,300 customers were powerless as the thermometers climbed to 112 degrees, another record. "There's a lore about how this is a temperate climate, and how we don't ever need air-conditioning, how we open the windows at night or go to the basement if we can," explained Vivek Shandas, a professor of climate adaptation at Portland State University (PSU), whose family bought its first AC unit to prepare for the heat dome. Open windows didn't help the workers at Starbucks, Burger King, and Chipotle, who called Oregon's OSHA office pleading for help in suffocating kitchens. Emergency rooms surged with patients, including victims of gun violence. One doctor suggested the heat pulled the trigger.

On Monday, June 28, the official air temperature at Portland International Airport reached 116 degrees. Multnomah County emergency services received about five hundred calls for medical assistance. Paramedics found men and women of all ages in the throes of heat illness. A thirty-nine-year-old man stammered through a heatstroke. An eighty-four-year-old woman was confused and alone. A customer who wandered into a service center slipped out of consciousness. A thirty-six-year-old woman lay face down on the sidewalk. That night, the heat finally broke, but it was too late for a fifty-seven-year-old woman who died at home, unable to run her air conditioner without tripping the circuit breaker. Daytime highs hovered in the nineties for two more days, prolonging the pain.

Not long after the heat dome lifted, scientists declared the event would have been "virtually impossible" without man-made climate change. Omega blocks, or jet stream undulations, occur naturally, but they are becoming bigger and moving slower as the planet warms and the temperature differential between the poles and the equator narrows. Those omega blocks in turn hold in atmospheric pressure systems that become heat domes. When the pressure builds, the heat compounds. In the past, an event like this happened once in a thousand years. But in the future, it could occur every decade, declared Zeke Hausfather, a prominent climate scientist.

Health officials began to sort through the human wreckage. Typically, emergency rooms in metro Portland admit around three heat-illness patients a year. But in six miserable days in 2021, they admitted nearly 160. Most summers, no one dies from heat, and in fact, Multnomah County had only recorded two such deaths since 2010. But under the heat dome, more than seventy lives were lost. More people died from high temperatures than from Covid-19 in the deadliest two weeks of the pandemic. More people died from heat in Multnomah County than in the entire state of Oregon in the previous twenty years.

But those were only the official numbers. Epidemiologists say excess deaths, which are the total number of mortalities above average for the time period, tell a truer story of the toll of heat. Excess deaths include not only those from heatstroke and hyperthermia, but also those that may be triggered by heat stress, like heart attacks and organ failures. In the Portland area, deaths from all causes doubled during the heat dome. The ghastly event may have taken 1,200 lives in the United States and Canada. If the temperate Pacific Northwest could suffer a spasm of lethal heat, then it seemed anywhere could.

Meteorologists and climate scientists looked to the sky for the causes, but the analysis of Vivek Shandas, the climate professor, was more down-to-earth. He wondered how the urban environment made the heat worse. Portland suffers a heat island effect that raises ambient temperatures in some neighborhoods as much as 20 degrees. Shandas has recorded and studied these local microclimates for more than a decade and offered his recommendations to cool them down. He knew the heat dome's temperatures would be higher on Portland's east side. But by how much, he couldn't guess.

On Monday, while most Portlanders were inside, Shandas and his eleven-year-old son left the cool refuge of their air-conditioned home to drive around the city with an infrared camera and a handheld thermometer, collecting temperature data. At around 2:00 P.M., they headed to a postindustrial district where brick warehouses had been converted to apartments and storefronts. Shandas stepped outside, and his eyes started burning. He stuck his thermometer in the air: 119 degrees, 3 degrees warmer than Portland's official air temperature, recorded at the airport. He pointed an infrared camera at the ground: 135 degrees on the surface.

They drove east, through neighborhoods that could have been anywhere in America, and passed strip malls and generic new apartment complexes surrounded by parking lots. Studies have shown that Portlanders can survive even the most extreme heat if their homes are up to the challenge. Heatproof homes need shaded windows, thick insulation, and effective cross-ventilation to pull in cool night air. But those essentials are not required by Oregon's building code, which meant the condos Shandas passed were uninhabitable hotboxes if their air conditioners were not working.

Father and son arrived in Lents, a historically neglected neighborhood that Portland has slated for redevelopment. The town center was supposed to be a bustling commercial corridor, but the streets were nearly deserted, except for a smattering of homeless tents near the freeway. The shadow of the overpass would have offered relief, but the city had dragged boulders underneath to deter the encampments. Of course, city officials weren't thinking about the heat when they did that. But that's part of the problem, Shandas believes. They never think about the heat.

Shandas pulled over next to a Planet Fitness on Ninety-Second Avenue and rolled down his window to measure the afternoon air. It was now a few minutes after 3:00 P.M. Shandas has recorded outrageous heat in the world's most sweltering cities: Doha, Hong Kong, and Hermosillo, Mexico. At 124 degrees, the air temperature here was the highest he had ever recorded anywhere. As the *Willamette Week* later declared, this was the hottest place in Portland. But you never would have known that by watching the news, or looking at your phone, which on that day reported a breezy 116 degrees.

Shandas got out to brave the conditions. He looked around and saw all the ingredients of urban heat. Asphalt: two four-lane roads intersecting near a twenty-thousand-square-foot parking lot. A cluster of tall and dark buildings, both absorbing solar radiation and impeding natural airflow. A few passing cars spewing waste heat from their combustion engines. No vegetation, except for some dry brown grass. And no shade, but for a few thin trees and the forbidden shadow of the overpass. For years, Shandas has warned officials that reckless urban development can raise local temperatures and amplify the effects of global climate change. But from the looks of this neighborhood, no one had listened. He pointed his thermal camera at the pavement. It was 180 degrees and could have melted the skin off his feet in a second. This neighborhood was an urban inferno.

There's not much that our bodies can do about 124-degree heat. Outside, three hours of exposure is fatal, even for young and healthy people. When the air is dry, total shade could represent the narrow boundary between life and death, so long as someone is lying down, completely still. The better option is to escape that weather altogether.

Shandas knew from his research that Portland's older, wealthier, and whiter neighborhoods, where grasses and trees outnumbered parking lots and cars, were cooler on hot days. He left Planet Fitness and zoomed across town. On the other side of the Willamette River, as the roads narrowed, he saw shrubby medians and majestic canopies that embowered the streets in shade. Shandas arrived in Nob Hill, a trendy northwest neighborhood. Just five miles from Lents, the air temperature was 111 degrees, 13 degrees cooler than the hottest place in Portland. Then he drove farther west to Willamette Heights, an exclusive hills neighborhood nestled in a verdant, five-thousand-acre park. There, the air temperature was 99 degrees, an astonishing 25-degree drop. For the young and healthy, such weather can be dangerous, but it is not an imminent threat.

Even before the official counts were released, Shandas knew the death tolls would be highest in neighborhoods like Lents, and lowest in those like Willamette Heights. Climate change does not just cause atmospheric disturbances. It does not just make big numbers race higher on thermometers. It also exposes urban inequality.

In the aftermath, Multnomah County health authorities identified the basic factors that risked the lives of Portlanders. Physical vulnerability was one. Seventy-eight percent of Portland-area decedents were elderly. As people age, their hearts weaken. Their sweat glands deteriorate, and that makes them overheat. Many decedents were already sick, and many did not have indoor cooling. Social isolation was a second risk factor. Seventy-one percent of those who perished lived alone. And urban heat was another. This is the issue that most concerns Shandas. The majority of casualties lived in areas that authorities identified as Portland's intra-urban heat islands. These neighborhoods had the scantest vegetative cover and the largest expanses of asphalt, and as a result, had the city's highest ambient temperatures. People suffered all over town, but 58 percent of deaths were in urban deserts like Lents. Global climate change had created the heat dome, but locally, the burden hadn't fallen equally on everyone.

The heat island effect is not a modern phenomenon. In fact, the tendency of paved areas to warm more than grassier and more treed surroundings was studied two hundred years ago by Luke Howard. The English chemist's book *The Climate of London,* published in three volumes between 1818 and 1833, was the first major work of urban climatology. Howard is best known for drawing and naming the clouds—cumulus, stratus, and cirrus—but he also noticed how city design and planning influenced the weather. Howard made daily recordings of the air temperature, pressure, and humidity at London's Royal Society and in the rural periphery. In the winter, the heart of the city was about 4 degrees warmer than the countryside. He attributed the warmth to the activities of heavy industry—foundries, factories, and steam engines—and home furnaces and fireplaces. "A temperature equal to that of spring is hence maintained in the depth of winter," he observed with only a slight exaggeration. During the day, the city was slightly cooler than the country, because the sun was shaded by a veil of smoke.

In the summer, it was a different story. There were fewer fires burning, and the skies were clearer. The sun shone more directly on London's brick buildings and cobblestone roads. These hard and dense materials have a higher heat capacity than grasses and soils, meaning they hold more thermal energy. Instead of evaporating the energy from the sun, as vegetation does, they store it and heat up, warming our feet through conduction, the air through convection, and even other buildings through outbound radiation. All this thermal activity meant London was warmer than the countryside, day and night. Only in the morning, when tall buildings shaded the narrow streets, did London offer an urban cool island effect.

Modern climate scientists quibble over the placement of Howard's thermometers and question the accuracy of his readings, but they agree on the fundamentals. Howard was among the first to identify the man-made causes of local warming and to diagnose them as the sources of elevated urban temperatures. Similar studies followed in Paris, Berlin, Vienna, and Manchester. By the mid-twentieth century, urban climatology was an established science. One such climatologist, Helmut Landsberg, was able to collect longitudinal meteorological data as the town of Columbia, Maryland, was hacked out of a forest. As green foliage was converted to concrete, asphalt, and dark roofs, the local albedo fell, which meant that less solar radiation was reflected to space, and more heat was stuck in the ground. Over a seven-year period, the surface temperature of the town center increased by 14 degrees. "The heat island is mitigated where daytime shade prevails," Landsberg observed, urging architects, engineers, and landscapers to preserve as many deciduous trees as possible. In infrared aerial photographs, "every tree stands out as a cool spot in the pool of paved and roofed area."

Scientists understood that urban heat impacted public health. The warm air belched by cars, stoves, and factories carried particles of dust, and the sun hardened those pollutants into murky and deadly air masses. In 1966, a catastrophic smog attack killed 168 New Yorkers over a three-day Thanksgiving weekend. To clean the air, Mayor John Lindsay established an air management network and partnered with university meteorologists to track atmospheric conditions. Lindsay's advocacy triggered greater awareness of the health problems of urban climates, and led to the passage of the nation's Clean Air Act.

But it was not until the 1970s that the threat of high temperatures came to the fore. At a National Institutes of Health conference, a former British navy surgeon named Frank Ellis presented an exhaustive analysis of two decades' worth of American hospital and coroner records. Ellis showed that summer heat waves were far deadlier than was realized because the stats didn't account for excess deaths. These deaths did not occur uniformly across cities. In 1972, an epidemiological analysis of a heat wave in St. Louis showed the fatality rate was higher in the concrete jungle downtown than in the leafy neighborhood near a large park just a few miles west. Yet these heat "epidemics," as Ellis called them, did not inspire government action. Unlike smog, this deadly urban weather was invisible. Moreover, it did not harm everyone—only the poor and vulnerable.

Shandas learned the importance of green design firsthand. He was raised in Bangalore, India, a busy city where there were more creatures than cars on the streets. Outside his window, the sidewalks were packed with food carts and stalls. Walkers, bikers, and drivers all shared the road, inching along together. Birds, squirrels, and other animals prowled the canopies of banyans and rain trees planted centuries earlier to temper the tropical heat. As a seven-year-old, Shandas came home from school to find his mother accosted by a troop of naughty monkeys rummaging through the kitchen cabinets. The monkeys had scaled a mango tree, scurried up the balcony, and busted through an open window. This was a lightbulb moment for Shandas. However human cities seem, they are still part of nature.

Shandas's family immigrated to America and settled in Santa Rosa, California, where they found their American dream in a split-level ranch house with a carport and a sycamore in the yard. The suburban look and feel were new to the boy. The streets were empty, and he was astonished by the quiet and isolation. No mango trees, no naughty monkeys. Only cars. It was the beginning of Shandas's lifelong interest in the origins and ethics of urban landscapes. Why are some neighborhoods green and lush, and others are bare and blanched? How do these environments embody a community's values? And who decides where the shade goes?

Portland is naturally coniferous, with hilly neighborhoods and high elevations that are cooler than those in the flats. But topographical features such as these are insufficient to account for the dramatic temperature differences Shandas has observed within the city. To explain why some neighborhoods are hotter and unhealthier than others, he looks to social history, as well as to nature. He is best known for groundbreaking research into the urban heat implications of redlining, work that reveals stunning correlations between the neighborhoods where racist appraisers discouraged investment in the twentieth century and the high thermal footprints of the twenty-first. Shandas and two other researchers compared satellite infrared images of more than a hundred American cities with the redlining maps published by the Home Owners' Loan Corporation in the 1930s. In every region and in almost every city, the neighborhoods that were redlined in the past had the highest land surface temperatures today. Subsequent research by the U.S. Forest Service confirmed these surfaces were hotter because there were fewer trees to shade them and because more of them were impervious to water and cooling moisture.

Shandas does not believe this is a mere correlation. He argues that residents of these neighborhoods were denied access to capital and could not buy property nor improve it with lawns and trees. Over time, the real-estate value of these neighborhoods plummeted. The low-cost land attracted large infrastructure that would have been more expensive and politically difficult to locate in wealthier neighborhoods. Urban planners slated the redlined neighborhoods for freeways and arterials. They designated the land for factories and thermally massive brick and concrete housing complexes. And when industry scaled back, developers swooped in to renew the neighborhoods with malls and residential infill surrounded by large surface parking lots. These are the reasons why some neighborhoods are hotter than others—they were planned and built that way. While the wealthy neighborhood's green space has been preserved and protected, the poor neighborhood's has been sacrificed. And there is nothing natural about that.

Shandas has pursued climate justice since 2009. That year, he published an article diagnosing the local environmental issues that were exacerbated by global climate change. At the time, decision-makers were developing carbon calculators for infrastructure projects and taxes for emissions. These worthy efforts only focused on mitigating climate change. They had little to offer people who already suffered from its effects. Shandas argued that it was time to start thinking about adaptation—how Portland could adjust to the climate-changed present and prepare for the hotter future. "If past climate-related events, like hurricanes and decade-long droughts provide an indication about what is to come, then adaptation is already needed on a massive scale," Shandas wrote. Those dramatic climate events were front-page news. But he believed the more insidious threat was the heat that killed more people than hurricanes, tornadoes, and floods combined. Shandas used an urban heat map developed by his university colleagues to show that higher temperatures were concentrated in Portland's poorest neighborhoods. If a heat wave ever struck, the people at risk would be on the east side.

In the years since, Shandas has offered adaptation solutions to Oregon authorities. The first kind of adaptation is through urban greening. The trees, grasses, and soils that grow naturally in Portland and other temperate cities are the strongest fighters against urban heat. The challenge is to find room for this vegetation in neighborhoods that are already built out. Green roofs are an appealing solution, but they are expensive and insufficient for urban-scale cooling. To meaningfully green a formerly redlined neighborhood, impervious surfaces like pavements and parking lots must be broken up and removed, and the land beneath must be reseeded. David Nowak, an emeritus senior scientist at the U.S. Forest Service, has even suggested tearing down vacant buildings and replacing the structures with green parks. Shandas made recommendations to Portland planners to unseal parking spots and leave more room in new developments for vegetation. Buildings are required to be earthquake resilient. Why can't they be required to reduce thermal stress? Streets are engineered to remove stormwater. Can't they be redesigned to repel the sun? Portland's urban planners use environmental overlays to protect rivers, streams, and forests from encroachment. Why not heat overlays to protect vulnerable people?

Shandas additionally recommends using urban form to fight heat, as the ancients did. In 2014, he and a geospatial researcher named Jackson Voelkel corralled a team of volunteers to drive around Portland and collect fifty thousand street-level temperature readings from dawn to dusk. They used predictive computer modeling to produce fine-grained, block-by-block heat maps that showed how a pleasant, 77-degree August afternoon on the leafy side of town was a bitter 95 degrees on the gray side. As expected, tree cover was a determining factor, but it turned out that tall buildings had an effect, too. Their data showed that downtown Portland had some of the city's coolest temperatures, likely because the streets were shaded by skyscrapers. The variance in building heights may have also promoted wind turbulence. This was surprising, because infrared satellite data usually shows urban cores as afternoon hot spots. From space, the neighborhoods appear to be dominated by the thermal footprints of their roofs. But as Shandas later said, "when you actually go down to the ground, where people are walking and life is happening, it turns out it's not the same signal."

Unfortunately, the cooling effect of afternoon shade was overshadowed by night heat. After the sun went down, the towers began warming the neighborhood by emitting the radiation they intercepted during the day, and by morning, downtown Portland was the hottest in the city. Although overnight effects can be mitigated by daytime reflective paints, Shandas offers another solution of thinner buildings to spare more room on the ground for heat-dissipating greenery. According to simulations conducted by Shandas and his colleagues, an urban corridor that has been rewilded and reforested can be 5 or 6 degrees cooler on a summer afternoon than the next one over. And as he observed under the heat dome, the differences can be greater when overall temperatures are extremely high. But instead of incorporating his suggestions of landscaped setbacks and limited surface parking into the zoning code, city planners have added them to an optional appendix for developers, where they have since been ignored. "None of it is mandatory," Shandas sighed, when we met at a coffee shop in Portland's Alberta neighborhood. "There's no carrots or sticks." The Lents commercial center, for instance, boasts no variance in building height to accelerate winds and whip up breezes, and the ground is flattened under sunbaked asphalt. Perhaps, when dangerous heat was rare, the microclimate effects of such planning decisions did not matter. Now, because of a changing global climate, they threaten public health.

The second kind of adaptation solution Shandas offers is not infrastructural but social. He is concerned by the isolation that proved fatal during heat waves in Chicago in 1995 and Paris in 2003, when thousands of elderly people died alone in their apartments. One reason is because confusion is a symptom of heat illness. When people are dehydrated or exhausted, they do not know they are in trouble. They need a friend or family member to check in on them and encourage them to chill in an air-conditioned library or other safe space. They need a neighbor to offer their cool basement to ride out the heat. And they need someone to take them to the hospital before their illness cascades into heatstroke. Shandas intuits the importance of social connections, and credits his community for helping him endure the tropical heat of Bangalore as a child. "The most obvious thing, on the top of my mind, is the neighborliness," he recalled with a smile, reminiscing about the apartment building where he grew up. "I could just go down a few floors, and hang out with people on the bottom floor, and they had a beautiful veranda where the wind would blow—so nice on the front porch—and I would hang out there, and they'd give me some cool drinks, and it was just lovely." His family did not rely only on a neighbor's veranda or iced sodas in their fridge, but public space as well. During a heat wave, an elderly uncle escaped his stifling apartment by sleeping for two nights on a patch of grass at a local park.

Shandas has encouraged Portland-area officials to be proactive. Rather than wait for a heat wave to arrive, he has urged them to educate vulnerable people now who live alone and don't speak English. They must know about the neighborhood parks, splash pads, and air-conditioned refuges where they can cool down. Shandas has suggested recruiting community groups with strong ties to spread the word. Alternatively, direct engagement could come in the form of a buddy system, or by designating block captains to check in on residents during heat waves. After all, neighbors are usually the first responders during a natural disaster. But along with his ideas about building codes and green infrastructure, Shandas said his pleas have fallen on deaf ears. Officially, Multnomah County has acknowledged the danger of heat waves. The government's climate plans have named heat islands as a top concern, along with their heightened health risks. But these plans are short on specific actions to alleviate the threats.

As the heat dome descended on the Pacific Northwest in 2021, local officials were unsure what to do. They had no early warning and response system for heat waves to activate. They didn't have the names and addresses of everyone who lived on heat islands. They didn't have a direct outreach system to check on them, and that wasn't something that could be pulled together in a few days. They raced to do what they could. The county published safety tips on social media in five languages and texted twelve thousand people, including new mothers and pregnant women, about the danger. The human services office called and texted more than four thousand clients, many in group homes. A nonprofit agency called managers of apartments and SROs and told them to direct their tenants to air-conditioned community rooms and lobbies. The homeless services office activated a network of mutual aid groups and nonprofits to hand out water bottles at encampments. An army of employees and volunteers fanned out to distribute cooling towels, electrolyte packets, and box fans. But the valiant effort was not enough.

Shandas was not surprised to learn that deaths were more numerous in four of the city's five hottest zip codes. The fatality rate in Lents was the highest of all. Shandas said their lives could have been saved. "I've been saying for fifteen years that these are the areas of town where we are going to see deaths," he told me, trembling with frustration. "This is the built environment and social vulnerability converging." And yet, he continued, "nobody gave it a second look. Only on Saturday, after the heat dome arrived, had they even started to open up cooling shelters. There was *nothing* done." If a hurricane or flood was a week away, he believed officials would not hesitate to act. But unlike those natural disasters, heat does not destroy property. It kills people.

Heat emergencies unmask the larger public health problems fueled by bad urban planning. The surge of deaths in sweltering neighborhoods but not shady ones attests to how land-use decisions have made some residents more vulnerable to heat's deleterious effects. Heat does not harm indiscriminately. Rather, it inflames existing conditions. Like the poor Angelenos in the ER, the Portlanders who perished in the heat dome had been poisoned by their surroundings, slowly and then quickly. Urban heat was a daily burden that weakened their bodies. Then, in an extreme event, it pushed many of them over the edge. Although the temperatures of the heat dome were unusual, the underlying risk factors were not. In fact, they were consistent with years of research about the causes of heat mortalities and morbidities. A heat dome can land anywhere, and when it does, it may well produce the same unequal outcomes that it did in Portland.

There are roughly eleven million Americans who live in formerly redlined neighborhoods, and the artificially elevated temperatures on these heat islands drive alarming health disparities. In a comprehensive national study, a team of doctors and epidemiologists calculated that a 10-degree increase in temperature can lead to a 24 percent spike in admissions to emergency departments. Ten degrees is a big jump—like spring giving way to summer. But in Portland, it's not even the difference between Nob Hill and Lents. In Richmond, Virginia, the city's warmest zip codes bake in July air that is 16 degrees hotter than the coolest ones. They also have the highest rates of heat-related ambulance calls and hospital and urgent care admissions, according to Jeremy Hoffman, a climate scientist who recorded heat islands there and in Baltimore and Washington, D.C., with Shandas.

Across the country, most residents stranded on redlined heat islands are Black and Latino, who as demographics are more likely to suffer the poor health conditions inflamed by heat, such as diabetes and cardiovascular disease. These diseases are linked to sedentary lifestyles, and shady, walkable streets can remedy them. A study of Sacramento neighborhoods linked a 10 percent boost in tree cover to statistically significant decreases in unhealthy weight, hypertension, and diabetes. In Miami, more trees were linked to fewer heart attacks among the elderly. But in cities across the country, the leafy, pleasant streets that encourage physical activity are usually located in white neighborhoods.

People who live on heat islands suffer high rates of asthma. Their lungs are scorched by nitrogen oxides belched from tailpipes and the fine particulate matter of acids and metals pumped out by diesel trucks and factories. On hot days, the blazing sunlight turns the polluted air into ozone, the chemical compound that blocks ultraviolet radiation in the atmosphere but causes lung problems down on Earth. As Frederick Law Olmsted suggested, parks can be a solution. Densely forested spaces not only filter particulate matter but also draw ozone and nitrogen dioxide out of the air. But there are few parks in neighborhoods with the highest asthma rates. On average, predominantly Black and Latino areas have 44 percent fewer green acres per person than do predominantly white neighborhoods, according to the Trust for Public Land, an environmental nonprofit. Instead of breathing fresh air, children in these neighborhoods reach for inhalers.

In previously redlined areas, the only place to play outside may be the school playground. But a typical public schoolyard is a grim asphalt blacktop that is cheaper and easier to maintain than a grassy field. Heat scientists believe these environments are dangerous to children, who are more vulnerable to heat than adults. They have a high skin to body mass ratio, which means they have more surface area exposed to sun; they don't sweat efficiently and rely on vasodilation to lose heat, which makes them weak and dizzy; they are also closer to the ground and intercept more heat wafting off the surface. And if they fall, their thin skin is more likely to be burned.

Although the future is uncertain, today's climate projection models make it easy to imagine what life could be like on the heat islands of 2050. Climate scientists believe the rate of warming could be double in cities than in the countryside. Over the last fifty years, as America's rural areas have warmed about a quarter-degree every decade, the biggest cities have warmed over a half degree. But that is just an average of the official temperatures recorded at single weather stations. The rise has likely been sharper in the neighborhoods that were fated to be smothered in sun-absorbent asphalt and bereft of shady trees. Unless we retrofit these neighborhoods for a hotter future, more people there could die what epidemiologists insist are preventable deaths.

And as heat waves grow longer, more frequent, and intense, the ill effects could strike in the spring and fall as well as summer. In a future that's too hot to play in, as physiologist Shawnda Morrison puts it, children won't have the same opportunities to run around outside, and their physical fitness will worsen, leaving them more vulnerable in the rare moments they do leave the house. And the public places designed for efficient automobile transportation and cheap maintenance could be deserted. The eerie emptiness of Portland streets during the heat dome could be more common in other cities where extreme heat could become chronic. Researchers have already identified Phoenix, Miami, and Los Angeles as cities that could be next.

Fortunately, we can do something about it. Shade can preserve the sidewalks that we need to stay healthy and to travel on in the city. In Phoenix, Arizona State University researchers used the wet-bulb globe temperature to determine that a twenty-minute "thermally comfortable route" that is safe to travel on 95 percent of summer afternoons needs only to be 30 percent shaded. A path like this could be shielded with a combination of "natural and engineered shade," such as metallic slat awnings and desert shrubs. Although this shade may not cool the air, it could indirectly turn down Phoenix's own severe urban heat island effect by encouraging more residents to walk instead of driving a car that contributes to the city's copious amount of waste heat. Through interventions like these, officials believe it is possible to make sidewalks feel cooler than they do today, even as global warming continues.

Shaded parks can be our refuges during heat waves. The air-cooling effects of dense vegetation are well known, and studies of urban parks in Vancouver and Sacramento have recorded ambient temperature reductions of 9 to 12 degrees under their canopies. Ideally, in tropical climates, the parks would resemble forests. But even in continental regions where sunlight is critical for winter comfort, a partially dimmed green space, dotted by dense patches of deciduous trees, could shade out the health risks of climate change. The humble shade sails that have no discernible effect on ambient temperatures can also preserve the accessibility of parks by cooling playground equipment. It takes only three seconds for a metal or plastic surface boiled by the summer sun to burn a toddler's skin and turn a family away. Shade eliminates that danger.

Shade can protect communities from heat, and when it is spread out, it can cool entire cities. According to the Georgia Tech professor Brian Stone, Jr., as a general rule of thumb, in large U.S. cities that already boast an average tree cover of 30 percent, a 10 percent boost could drop summer temperatures by 2 degrees, erasing the global warming predicted by the middle of the century. Yet cities can aim higher. Research conducted by the Los Angeles Urban Cooling Collaborative, an interdisciplinary group of university and nonprofit climate scientists, projected that the temperatures of fatal heat waves in L.A. could be at least 3 degrees lower if the county was 40 percent forested. In combination with other adaptation measures, like reflective roofs and roads, L.A. could change its weather for the better, and shift humid, oppressive air masses to more benign ones, and bone-dry heat to milder warmth. This may not sound like much cooling, but a drop like that could save at least one in four lives currently lost to heat waves, mostly in low-income communities. According to the scientists, an aggressive climate adaptation effort such as this could delay the local effect of global warming for a hundred years. Even a more modest reduction in temperature could improve neighborhood health for decades to come.

A few degrees off a summer high may not fully close L.A.'s climate gap, and according to Shandas's research, it would only make a dent in Portland's. Nor would it avert all the dangers of climate change, such as unpredictable weather and life-threatening heat domes. But doing nothing should not be an option. Through forward-thinking urban planning, communities can be safer from heat than they are today. Their futures do not have to be determined by the poor planning decisions of the recent past. We already have the technology we need. Now our leaders have to find the political will to use it.

Part III

The Future of Shade

Chapter 7

Shelter from the Sun: Architects Revive the Dream of the Naturally Cool Home

As Portland's Lents neighborhood burned under the 2021 heat dome, Jeff Stern, a calm and brainy architect, watched the indoor thermometer in his Beaumont-Wilshire living room. Sunlight yearned to rake across the concrete floors and clean white walls of his loft-style home. On Saturday morning, the official temperature at the airport cracked three digits. On Sunday, it soared to 112 degrees. On Monday, to a breathtaking 116 degrees. Yet the temperature in Stern's house never ticked above 84. "Light sweating mode," Stern laconically called the heat indoors, as he and his wife Karen sauntered around the house in shorts instead of their usual garb of jeans and cardigans. They weren't too bothered, he said. Nor was the dog, a German shepherd mix named Dee.

Eighty-four degrees inside isn't what most consider comfortable, but it is safe. Stern doesn't live in a leafy enclave like Willamette Heights, in the cool shadow of a natural forest. His east side neighborhood isn't much shadier than any other in Portland. Stern's house doesn't even have air-conditioning. What it does have is a superior ability to shut out the sun and bottle up cool night air. Stern lives in a passive house, a new kind of building governed by old design philosophies. These homes consume only a fraction of the energy that is typically needed to stay warm in the winter and cool in the summer, and they achieve this superior performance by combining the ancient methods that once dictated the architecture and urban design of arid lands with the solar control techniques that were advanced in the mid-twentieth century by architects like Victor and Aladar Olgyay.

As pressure mounts to use energy more efficiently to slow the burning of fossil fuels that drive climate change, passive houses are rising in popularity. But by reviving these cool old ideas, these low-energy homes also introduce a chilling new one. The same features that make passive houses sustainable also make them more resilient in extreme weather. In a world where the climate is becoming more frightening and unpredictable, and rising heat threatens to surpass the limits of human survivability, passive houses can be safe havens. To survive a climate-changed world, we may all need to live in passive houses that have the ability to shut it out.

In the 1980s, as Europe recovered from a decade of oil shortages and fluctuating energy prices, a German physicist named Wolfgang Feist and a Swedish building engineer named Bo Adamson struck up a conversation about conservation. Adamson had returned from southern China, where heating fuel was banned and residents had no choice but to find other ways to endure the winter. Could the same practices be applied in Europe, where the winters were even colder? It was worth a shot.

Since the Industrial Revolution spread access to affordable energy, indoor climate control has been mostly mechanical, driven by the rise of so-called active machines like boilers and air conditioners. But for most of human history, the principles of passive cooling and heating were the foundation of building design in the world's most challenging climates, like the deserts of the Middle East, North Africa, and the

American Southwest, and in the frigid cold of Canada and Scandinavia. Passive houses revive and perfect the techniques that were once used to ensure indoor comfort without any energy at all. These dwellings can be big or small, new or renovated, and they don't even have to be houses. Since the early 2000s, apartments, offices, businesses, and hotels have been certified as passive houses in Europe, the United States, and China. Theoretically, any American building can be a passive one, so long as the architects or occupants prove it meets a rigorous energy quota. To officially qualify, a passive house must not consume more than fifteen kilowatt-hours of electricity to heat or cool every square meter of living space over the course of a year according to the Germany-based Passive House Institute, one of two international certifying agencies. That's roughly a quarter of the rate of a typical American home, estimated Ken Levenson of the Passive House Network, which promotes the building method in the United States.

Stern's home, which he designed himself, easily met that standard. From the street, the two-story, 1,965-square-foot house, which he completed in 2013, looks clean and modern: two conjoined boxes, clad in brown cedar and accented with Mondrianesque red stripes. The mid-century modern aesthetic camouflages the traditional thermal design. A principal consideration is solar orientation. Stern's house faces south, admitting sun for natural warmth in Portland's cool seasons. Another is insulation: lots of it, easily twice the code minimum, to emulate the thick walls of a mudbrick hut and slow heat transfers. The foundation, walls, and roof all fit together to create an airtight seal. The triple-pane windows are 40 percent more efficient than standard glass and are fitted precisely in the walls to eliminate leaks. Stern's crews sealed every crack in the interior sheathing with industrial tape, a painstaking process. Instead of unwanted drafts, the air moves in deliberate currents. Natural ventilation is key: Both wings of the house have high ceilings, and the living room soars to seventeen feet, allowing hot air to rise. In the summer, when the sun goes down, Stern opens the windows to bring in the evening breeze. "Night dumping," he calls this process of natural air-conditioning.

Not long after he wakes in the morning, he scoots around the house, closing the windows to seal in the cool.

Windows aren't just our visual connection to the outside world. They're our thermal connection, too. They are loose and leaky, and the glass itself is a terrible insulator. Building engineers call windows the holes in the building envelope, because even when they are closed, they are still open to hot and cold air. Passive house designers try to overcome those defects with better frames, tight seals, and additional panes, but even the thickest, most sun-repellent glass allows more heat than an opaque wall. Stern punched a few small windows through the north, east, and west walls for ambient light and air circulation. On the south side, his walls are practically transparent, dominated by sweeping, fifteen-foot-wide rectangular windows and sliding glass patio doors.

As the modernists discovered in the twentieth century, all that glass can turn dangerous in the summer. Now, in the climate-changed twenty-first, architects are learning that lesson all over again. In Manchester, Edinburgh, and Glasgow, cities that once had mild, foggy weather, residents of glassy high-rises where windows are sealed shut for wind safety are getting roasted on sunny days. In San Francisco, where notoriously cold summers are becoming balmier, condo owners have sued their architects and developers for damages, claiming glass has rendered their sun-facing apartments uninhabitable. Large, unshaded windows in a typical home can easily account for half of summer cooling loads, according to Stephen Selkowitz, a building scientist at the U.S. Department of Energy's Lawrence Berkeley National Laboratory (LBNL). Stern's sun-blasted living room would have been painful in the heat dome, were it not for his home's other passive design features.

The developers of one San Francisco condo defended the excess of glass by pointing out that the building was code compliant. But that's the problem. Architects, engineers, and developers use past conditions to determine if new buildings will be safe in the future. They use historic weather data to formulate their cooling systems, insulation levels, and glazing. Now, a changing climate calls the utility of the old data into question. This leaves them with a whole new set of design challenges. When architects work in climates that still have dreary winters, how will they ensure that a brilliant sunlit interior doesn't become a hotbox during an unanticipated freak heat dome? How will they strike a balance between hot and cold? Stern, and other passive house architects, may have a solution.

Contrary to what Emma Stone's and Nathan Fielder's obnoxious characters say on *The Curse,* the TV show that mocks sustainable architecture, passive houses are allowed to have air-conditioning. The idea is simply that the systems don't need to be so large, or run so often, as they would in a drafty and poorly insulated building. A few quick blasts go a long way. When Stern broke ground on his passive house in 2012, mechanical cooling was still somewhat unusual in Portland by national standards; at the time, only 41 percent of homes were air-conditioned. Because of the insulation, Stern could have installed AC, used it infrequently, and stayed comfortable while meeting the passive house energy quota. Instead, he bought window shades.

A window shade's ability to block the sun and cool a room depends on a few factors. First, the color and material. Since the days of the Olgyays, polished aluminum has been a preferred option, reflecting more solar radiation than a white finish. Second, the location of the device. This is critical. As Le Corbusier learned, when sunlight passes through a window, it warms interior surfaces, including blinds and curtains, and the radiant heat can't escape. To eliminate solar heat gain, shades must be installed outside. The Olgyays had a complex and practically unusable method for constructing the perfect exterior shading device, but short of that, they recommended venetian blinds, like the ones Stern would install over his windows.

When closed, exterior venetian blinds keep Jeff Stern's house cool under high temperatures without air-conditioning.

Because solar control is a lost art in America, Stern could not find those shades at the local Home Depot. Actually, he could hardly find exterior shades in the United States at all. Although building scientists remain unanimous in recommending them for smaller cooling loads, cheap and abundant energy has made them a niche product here. After some digging, the architect landed on remote-controlled aluminum blinds made in Austria, a country where air-conditioning is still uncommon. A set for his three large windows cost an eye-watering $10,000. "As much, or more, than installing air-conditioning," he admitted. "There's definitely a premium for it." But like an auto enthusiast who imports a classic ride for the craftsmanship, the aesthetic, the cred, and above all the superior performance, Stern could not accept anything less. With louvered blinds, Stern could have sunlight when he wanted it and shade when he didn't.

He is not the only passive house architect who relies on such blinds to solve the design challenge of sunlight. When Wolfgang Feist commissioned the first *Passivhaus,* a superinsulated, three-story brick building for four families in Darmstadt, Germany, he oriented the building exactly to the cardinal directions, and instructed his architects to locate generous glass faces on the southern exposure. (East and west exposures receive direct light only in the morning and afternoon, respectively.) In the chilly winter, his design worked as intended. The windows admitted loads of solar radiation, and the natural warmth reduced the demand for heating energy in his seventeen-hundred-square-foot home by an astounding 90 percent compared to a typical build. But over time, the *Passivhaus* stopped working as intended. In the early 2000s, as Europe was convulsed by the first of many vicious heat waves, the glass house became a greenhouse. The key to cooling down was shade, and Feist needed to get it right. Too much would end the winter warmth, and too little would be ineffective.

Feist's windows are recessed into thick walls that hang over the glass like furrowed brows. On the lower floors, the windows are also located below shallow terrace balconies. On the summer solstice, when the sun is highest in the sky, these overhead shelters make enough shade. But air temperatures do not peak in late June. In Germany, the United States, and other mid-latitude countries, temperatures rise through July and August, even as the sun's angle declines. And these days, September and October can bring an infernal combination of high temperatures and lower-angle sun. In these months, when sunbeams strike vertical surfaces more directly, a horizontal shade like a lintel or a balcony is not effective. To prevent overheating, Feist correctly deduced that he needed vertical shade and turned to an exterior blind as the simple, elegant, and versatile solution.

When it comes to sun protection, exterior venetian blinds can't be beat. When the slats are open, light can still come through but most of the sun's heat is gone. Close the slats, and the blind becomes opaque: near-total zero-solar 100 percent shade. Even in homes and offices that are air-conditioned, the energy savings are significant. The shade greatly reduces AC consumption. On a hot, sunny day, venetians installed outside south-facing windows can cut a London office's cooling load by 37 to 40 percent, the United Kingdom's National Energy Foundation estimates. In sunnier climates like California, inch-wide matte white blinds reduce AC use by 78 to 94 percent in the afternoon compared to conventional interior shades, according to an LBNL study. Between drawing the blinds in the morning and retracting them to open the windows and store cool air at night, Feist's home remained comfy. When Feist installed arrays of photovoltaic panels in 2016, he tilted one set over a third-story window, as a small solar awning, protecting the glass from the sun's rays while converting them into usable energy.

Passive house architects use computing tools to calculate the energy use implication of every design decision they make. The software strongly discourages decorative elements, because every additional cantilever or corner or dormer creates another surface that needs to be insulated, and another opportunity to lose energy to the outdoors. For that reason, most passive houses look boxy, like Stern's. Shades prove the exception to the rule of austerity. Passive house architects let their imaginations run wild with solutions to solar gain. Southern windows may be sheltered under an overhanging roof or an outrigged metal slat, like an awning bracket without a fabric. The glass might be hidden behind rows of louvers or shutters. The edges of eastern and western windows could be flanked by vertical fins, flying off the building like flags in the wind. Some windows are shrouded behind metal boxes or recessed into sculpted façades. Other passive houses hardly need window shades at all. Instead of intercepting the sun before it strikes glass, some passive house architects simply use less glass to begin with, about half of what's recommended in standard building codes.

The first day under the heat dome, Stern awoke in his second-story bedroom. Downstairs, the morning sun was edging into the living room. The architect watched his weather app and the indoor thermometer. As the temperature outside neared the temperature inside, he began to close the windows, locking in the overnight cool. In the living room, Stern held down a wall switch, activating an internal motor that hummed distantly, like a washing machine in another room. Thin aluminum blades, about an inch deep, rolled out of the window heads and descended the glass like slow guillotines. The sun glared on the metal and reflected softly through the windows. Stern pressed again and the slats slammed down, plunging the room into darkness. Daylight seeped into the living room from the other side of the house. Outside, the city was withering in desert heat. But inside, Stern said, the hazy gray light reminded him of cloudy skies on a rainy day.

Americans are too reliant on air-conditioning, an energy-hungry machine that worsens climate change. Many of us live in homes that would be stuffy and uncomfortable without AC, and in extreme conditions turn dangerous. We have a solution to this problem in the form of passive housing. Yet at present, it accounts for less than 1 percent of new American construction in the past decade. Why isn't it more popular?

Passive houses embody what is known in energy efficiency circles as the split incentive problem, or an investment whose benefits do not accrue to the investor. Jeff Stern had a unique opportunity to build his own low-energy home and eventually recoup the costs of his up-front investments in the form of smaller utility bills. But the vast majority of American homes are built by developers and sold to someone else, and the vast majority of those developers are not financially incentivized to minimize their future occupants' need for cooling, heating, and other utilities. In fact, the largest developers resist efforts to increase the home insulation needed to improve energy efficiency.

Stern experiences the split incentive problem when he designs other peoples' homes. Years ago, he was hired to design a four-story, twenty-three-unit apartment building for a developer in Portland. On the southern exposures, Stern called for wood shutters to slide on a track outside the windows, like a barn door. For most of the year, the shades would be out of the way, but in the summer, he imagined residents could crank open their casement windows, reach outside, and slide the shutters across the glass. But early on in construction, the developers began to worry about the durability of the external shades and backed out. What the tenants got instead were window ports to install their own AC units. Instead of paying more up front to save energy, the developer opted to pass the costs of cooling on to the future tenants.

From a business perspective, the decision of developers to cut corners and not to invest in shading and other energy-saving measures makes sense. But the impact of decisions they make to protect their individual bottom lines affects all of us in the future, in the form of more and more energy that will be needed to keep those buildings cooler and cooler as temperatures continue to rise. It is this understandable self-interest that has prevented passive housing from reaching the masses.

To cut through this knot, Americans need stronger building codes that mandate higher energy-efficiency standards, and developers need stronger financial incentives to build climate-resilient homes. We have reasons to be optimistic. Since Stern finished his passive house in

2013, such policies and incentives have boosted passive house construction in states such as New York, Massachusetts, Pennsylvania, and Colorado, where political leaders are committed to reducing the carbon emissions of the building sector. To achieve these goals, they have not only beefed up their building codes but also awarded tax breaks and direct cash incentives to homebuilders who meet passive standards. Because of their unusual requirements, a passive house, apartment, or office can cost 3 to 5 percent more to build than a conventional structure. These states are offsetting the cost of that construction through government support.

Currently, there are about sixteen thousand passive house apartments in the works nationwide. If every state in America strongly committed to reducing energy and increasing climate resilience, then passive housing could one day be as popular in the United States as it is in the European Union, where member countries are commanded to reach zero emissions in new buildings by 2030 and in existing ones by 2050. This mandate has made passive house construction all but essential in many countries. The Continent boasts many thousands more passive houses, apartments, and offices than the United States, and while most are new builds, some are also retrofitted old stock.

"I move into a building, and it's marketed as a luxury building," said Alexandria Ocasio-Cortez, the congresswoman who represents New York City's outer boroughs in Washington, D.C. "It's an efficient building, it's clean, it has public space, it has a rooftop garden." A few months after she first introduced a Green New Deal resolution that set a long-term goal of sustainable and affordable housing for every American, Ocasio-Cortez returned to Queens to open eight stories of passive house apartments for the elderly, many of whom were formerly homeless. "I do a tour with one of the seniors," she continued, "and it looks just like my apartment." She was stunned. Her visit occasioned a question: How could safe, dignified, and environmentally responsible housing be available for everyone? In part, through passive housing. The technical prerequisites for the world that she and others fight for are already here.

The Deceptive Promises of Green Designs

Or are they? "I'm not sold on the passive house as the model," said the architectural historian Daniel Barber. "It seems to serve some purposes, but it's very heavy."

I expected Barber, the author of *Modern Architecture and Climate: Design Before Air Conditioning,* to laud passive construction. His 2020 book celebrates postwar architects like the Olgyays who relied on building design to create natural cooling. But when I reached him over Zoom in Germany, where he was a research fellow at Heidelberg University's Centre for Apocalyptic and Post-Apocalyptic Studies, he was less than enthused.

Barber explained that passive houses are still high-carbon architecture. Although the operation of the building doesn't consume much energy, huge amounts of fossil fuels are needed to produce the insulation that holds in the cool. Carbon is embodied in the very structure of the building. Spray foam is a common insulation material in passive houses, and the additives used to make it are themselves greenhouse gases. Some of these gases trap easily a thousand times more heat than carbon dioxide. Barber's concern is that the emissions associated with the production of insulation, and the construction of passive houses, will cancel any savings on air-conditioning down the line. Some studies are bearing him out.

A different version of sustainable architecture does not require insulation but greenness. Building roofs can be covered in grasses, and walls can be scaled by vines and moss, to absorb solar radiation and dissipate the heat. Balconies can be gardens that accomplish the same thing. The green architecture trend kicked off in the 1960s and 1970s, when a handful of stunning buildings arose in the United States, the Netherlands, Austria, and Japan, stoking the fantastic dream that we could all be tenants of the trees. In the twenty-first century, an image of two tree-covered apartment towers in Milan has become the new visual shorthand for sustainability, a real-life "nature is healing" meme. Buildings account for around 40 percent of global greenhouse gas emissions. Rather than add these gases to the atmosphere, these green buildings suggest that it might be possible to remove them.

On a balmy summer afternoon, I toured those tree towers. The two residential skyscrapers, known as the Bosco Verticale, stand at nineteen and twenty-seven stories, and they opened in 2014 to international acclaim. What could be more luxurious, and more fantastic, than living amid the foliage that is so scarce on the ground below? Anastasia Kucherova, a project manager at Stefano Boeri Architetti, the studio that designed the towers, noted this love of greenery. "People would rather live surrounded by nature than have golden bathrooms," she claimed, as we stood on a spacious fourteenth-floor balcony, surrounded by flowering bushes and fragrant aromatics. This wasn't like having a few potted plants on the windowsill. The sight of city streets through a thicket of leaves and branches took me back to childhood. I remembered what it felt like to climb a tree. At the time, the building's smallest one-bedroom units sold at around 600,000 euros, among the highest per-meter prices on the Continent. The largest units sold for more than 4 million.

From the ground, the towers resemble two columnar arborvitaes, receding into the bushy umbrage. The actual trees are housed in hundreds of heavy-duty, concrete planter boxes, built into spacious balconies that are cantilevered off the façade. To withstand the heavy winds in the sky, the trunks are belted and fastened to steel cables plunging from the terrace above. Miraculously, most of the trees have survived, thanks to an ingenious engineering and landscape team, and an acrobatic crew of aerial arborists who rappel down the building face, tending to the gardens as they hang in a bosun's chair. When trees and plants do die, new ones are lowered into the planters by roof-mounted cranes.

There are good reasons to cover skyscrapers in trees. Butterflies and bees make homes here, restoring biodiversity to a city lacking animal and insect life. (Boeri, the architect, claims the towers brought swallows to Milan.) Then there's the calm, peaceful swaying of the leaves and the fresh fragrances. Mainly, the greenery has environmental benefits. The foliage screens airborne pollutants, dust particles, and noise, and evaporates solar radiation. The shade might be the biggest benefit of all. Heat-tolerant evergreens from the Mediterranean, like olive trees and holly oaks, block the sun on the southern and western façades, and on the hottest day of the year, reduce solar heating at least 40 percent, according to a study by the Council on Tall Buildings and Urban Habitat. That's especially important in Italy, where authorities limit air-conditioning and set minimum indoor temperatures to save energy. By extension, the shade benefits everyone else in Milan, because the foliage reduces the amount of waste heat the towers' air conditioners would otherwise expel into the city.

But to call the Bosco Verticale sustainable would be wrong. Skyscrapers have massive carbon footprints. They need huge steel columns to resist the force of high winds, and steel production relies on coal to melt the iron ore. They need a prodigious tonnage of concrete to build the foundation and internal core and floors. Concrete is an environmental nightmare responsible for about 8 percent of all human CO_2 emissions. Besides the fossil fuels burned to power the manufacturing, the chemical process of mixing cement releases carbon, too. Every metric ton of concrete adds another 1,370 pounds of CO_2 to the atmosphere. Once built, the towers suck up lots of energy to deliver water to the upper stories and to heat and cool so many rooms, even taking the shade into account.

The Bosco Verticale requires more than the typical amount of concrete to support all those thousand-pound trees. Although the studio never released figures, Boeri estimated the overall surface area of the eleven-foot-long cantilevered balconies and their deep planters at about ninety-six thousand square feet. "That is a lot of concrete, with a big carbon footprint," wrote architecture critic Lloyd Alter. Because a mature tree can absorb less than fifty pounds of carbon per year, it might take the vertical forest's eight hundred trees "a thousand years to pay back the carbon debt of the planters they sit in," he sniffed.

From my leafy perch in the Bosco, I peered down to the old low-rise courtyard apartments (*case di ringhiera*) clinging to the edge of the neighborhood. I could see the window shutters that urban Italians traditionally used to filter the sun, and the iron gallery balconies where they took in cool night air together. Barber, and other architecture critics, believe retrofitting existing buildings such as these may be a more sustainable path than building more new ones. No matter how much power the Bosco Verticale saves, it may never offset the carbon emissions needed to construct it in the first place. No amount of trees may ever make it green.

The green fetish distracts us from solutions that are simpler, cheaper, and more sustainable—and when it comes to fighting heat, more effective. Maybe in the future, green architecture won't be green at all, but beige, brown, or red, like the mudbricks that some architects in Niger and Burkina Faso use to build homes, schools, and mosques. Like the adobe homes of the American Southwest, these buildings are constructed onsite, using the material in the earth around them.

Or maybe green architecture is white, like the picturesque, limewashed hill towns of the Mediterranean. If mudbrick's superpower is the ability to slow the intrusion of solar radiation, white paint's calling card is high albedo, or its ability to reflect the radiation away. A freshly painted white roof might bounce anywhere from 81 to 88 percent of the sun's energy back into space, absorbing only the remaining fraction. By comparison, a black asphalt shingle roof will have the opposite effect, reflecting only 4 percent of sunlight and converting the rest of the energy into heat. On a hot and sunny day, a white roof might warm to 100 degrees, but a black roof will soar to 150. In the dense urban centers of hot and energy-poor countries like Sierra Leone, India, and Indonesia, where space for trees is at a premium, white roofs are a no-brainer.

In America, reflective roofs have been required in building codes since the early 2000s. Some of them aren't even white, but rather, painted with specially engineered tan, salmon, and green pigments that absorb some of sunlight's visible spectrum while deflecting the infrared. This tech is getting better all the time. University scientists have tinkered with a super-reflective film that creates "passive day-time radiative cooling." Besides reflecting sunlight, this film also ab-sorbs heat from other surfaces and shoots that energy into space, too. The astonishing result is that a roof coated in the film can be about 8 degrees cooler than the air above it, which is a strange, uncanny sensation, like a touch of shade in full sun. If the technology ever scales up, it could be immensely powerful.

Barber, the architectural historian, has a different idea. He does not believe that gradual technical improvement—whether through insulation, greening, or reflectivity—will end global warming. After all, the more efficient a technology becomes, the more we tend to use it. And so far, any improvements in air-conditioning technology have been overwhelmed by the growth in emissions rising from global demand. Instead, Barber has suggested a life "after comfort." Maybe architecture should not become thicker and more protective, and maybe we should not retreat into proverbial climate fortresses. Maybe architecture should become more open and porous, so we can learn to live with heat, instead of seeking the most efficient method of avoiding or obliterating it.

Barber's provocation is convincing. Studies from around the world demonstrate the physical and psychological elasticity of human com-fort. A determining factor seems to be choice. Those who are forced to deal with heat have proven more tolerant than those who can es-cape it. For instance, in the United States, building engineers have determined that the ideal indoor air temperature range (comfort zone) in the summer is roughly between 73 and 78 degrees, condi-tions that can often only be achieved through mechanical cooling. But in impoverished cities, like Ouagadougou, Burkina Faso, where there is no such option for artificial chill, similar studies have deter-mined that the Burkinabe prefer warmer air at around 86 degrees. Few Americans would ever call that comfortable.

To transcend the American standard of comfort, we might look to architects who work in that hot and dry Sahelian country, such as Francis Kéré, who hails from the remote village of Gando. As a seven-year-old child, Kéré left his family to attend primary school in a nearby city. The schoolroom was a concrete bunker with no electricity, and except for a few small windows, no light. He was one of a hundred students crammed in a dark, humid room, unable to learn. Eventually, Kéré left Burkina Faso to study carpentry and architecture in Germany and vowed to return home to improve the conditions of other Burkinabe. Kéré did not aspire to move rural children into brightly lit and air-conditioned schoolrooms. Rather, he wanted to use the materials readily available in Gando to design an environment that was modestly more conducive to learning. Without much available electricity, the challenge was formidable. How could he let in air, but not dust? How could he keep rooms cool, but not too dark to read a blackboard or write at a desk? And how could he strip the heat from the sun while preserving its light?

In 2000, Kéré began to build a single-story, rectangular primary school in Gando that would comprise a row of three classrooms interposed with outdoor courtyards and porches. The structure was designed to be oriented east to west, to limit solar heating of the side walls. The first step was to make the bricks. Kéré fundraised for Gando villagers to compress clay onsite and leaven the bricks with concrete. The clay offered insulation, and the concrete offered durability and rain resistance. The next step was to design the roof. Corrugated metal roofs are common in Africa because they are waterproof, but the metal is also an effective conductor of solar radiation and blasts heat inside. Kéré devised an ingenious solution to detach the roof and suspend it above the building on metal trusses, which dissipates the heat. The parasol roof hangs over the walls, protecting the bricks from the rain. It also throws a large shadow over the whole school building, shading the ceiling, the façades, the internal courtyards, and the humble outdoor terraces. The roof system is signature Kéré and has since been emulated across the African continent.

In Burkina Faso, students at the Gando Primary School gather in the shade of the building's overhanging roof.

The Gando school, and an extension finished in 2008, also sport metal shutters in the slim, vertical windows Kéré punched through the bricks. Besides solar control, the shutters create air circulation. These windows have no glass, and when desert air is sucked through the louvers, it accelerates and naturally fans the students' skin. The currents of wind carrying their body heat rise and escape through holes in the ceiling. Between the thick insulation offered by the mud-bricks, the ventilation created by the windows and ceiling, and the shade of the parasol roof, Kéré says his buildings stay naturally comfortable until late afternoon. At that time, students repair to the shady outdoor niches and courtyards, just as their parents come together in the afternoon in the shade of the town palaver tree.

Kéré creates sophisticated passive cooling systems that are not turned on with the push of a button. The architect's buildings, which also include teacher housing, medical centers, and secondary schools across Burkina Faso, do not beat the heat through air-conditioning, or defend against it through airtight enclosures. Instead, the structures use the natural elements of shade and wind to temper a hot desert environment and create a tolerable level of comfort. But as Kéré himself has said, these systems are complicated and require engaged users to succeed. In Gando, for example, the schoolteachers must know when to open windows to let cool air in, and they must know when to close them to block desert sun. They must know when drowsy outdoor warmth eclipses stifling indoor heat, and they must plan their lessons around this daily thermal rhythm. At another school, Kéré has experimented with evaporative cooling, leaving niches for buckets of water under the windows, so the wind can blow much-needed moisture into the dry classrooms. The staff must clean and refill the vessels regularly. It is not enough to design a passively cooled building. Occupants must know how to actively operate it.

Studies have shown that thermal awareness is key to life without air-conditioning. Hyderabad, India, shares a hot and dry climate with Ouagadougou. In 2008, building researchers surveyed Hyderabad residents and discovered that people who lacked mechanical cooling tolerated higher temperatures better than those who relied on AC. Some respondents who could not afford cooling said they were comfortable in 91-degree heat. This temperature does not pose a physiological threat to healthy people inside. To avoid discomfort, they opened their windows to bring in breezes. They turned off their lights to minimize waste heat and worked in the glow of the window. And they turned off metabolic heat by napping. A study from Ghadames, Libya, that compared comfort in naturally ventilated and air-conditioned homes came to a similar conclusion. The more time one spends coping with heat, the more bearable it becomes. Should more Americans ever hope to kick their addiction to cooling, it will first be critical for them to reacquire the thermal knowledge that we once relied upon to behaviorally thermoregulate.

Some Burkinabe who live and work in Kéré's buildings claim to be tougher. Even when the temperature outside is 104 degrees, they say they don't need air-conditioning. Thick walls, natural ventilation, and shade are all the cooling they need. Of course, this is not a choice. Burkina Faso is an impoverished country. Less than a fifth of the population has any electricity at all, much less power for air-conditioning. Their environmental blamelessness is a result of enduring poverty, caused in part by drought conditions that have made farming even more challenging. They are already suffering from extreme man-made heat waves that will worsen in the decades to come. But it is also true that Burkina Faso, which accounts for one-fiftieth of 1 percent of global emissions, is not at fault for climate change. If we all lived a bit more like the Burkinabe and learned how to cope with heat, then we would not consume so much of the cooling energy that continues to warm the planet.

Perhaps, if we can find a way to live without air-conditioning and all the other creature comforts, it will be because we have first trained our bodies to accept and tolerate more heat than we currently do. Comparative research conducted in Marrakech and Phoenix confirms that comfort is relative. In similar weather, Marrakech residents are twice as likely to feel comfortable as those in Phoenix, in part because they are more likely to live in naturally ventilated homes and spend more time outdoors. More exposure to the elements calibrated their thermal expectations and helped them tolerate environments that air-conditioned Americans prefer to avoid. Moreover, physiological research confirms that humans can improve their autonomic thermoregulation to better handle the physical stress of high heat. Our bodies adapt after two weeks of gradually more outdoor exercise. We sweat more, it comes on faster, and it happens in new places—not just our heads and backs, but on our arms and legs, too. As Barber would say, we adapt to heat by making ourselves uncomfortable.

The architectural historian has imagined a situation where the energy-abundant Americans who fuel climate change gradually begin to transfer some of their thermal wealth to countries like Burkina Faso, in what he calls "comfort reparations." In this scheme, which would require something like a global energy quota, Americans would endure less comfort, and reduce their wanton use of AC, so that countries on the front lines of climate change could begin to use AC to survive life-threatening heat. Americans would only be allowed to return to our previous air-conditioned life when we have finally gotten off fossil fuels and are on a totally clean grid. Among other things, this utopian vision would require us to sweat more. It would require us to slow down. And it very likely would force us to appreciate the shade of a tree a little bit more than we do today. "When it's 140 degrees out, I hope to God I have an air conditioner, and that you do, too," Barber has said. "But when it's 85 out, please just open the window."

Chapter 8

A Different Light: Innovating Shade for a Warmer World

For a moment, it was the world's most infamous shade. On a dreary May morning in 2023, Los Angeles city officials convened the press at a shabby sidewalk on Third Street, a major thoroughfare out of downtown, to unveil what they believed to be an exciting improvement to the experience of riding the bus: an innovative shade structure called La Sombrita.

This was the city's attempt at addressing riders' perpetual lament that thousands of bus stops are too damn hot, and specifically, at making the bus system work better for women, who often have needs that men don't. They're more likely to travel with children in strollers and elderly relatives in wheelchairs and need space and shelter to accommodate them. They're more likely to travel in the middle of the day when the waits may be longer and the temperatures higher. And they're more likely to feel unsafe when they travel at night.

When you hear the words "innovative shade," what comes to mind? Maybe you're imagining a slick shelter made of photochromic glass, dark in the day and translucent at night. Perhaps an elegant, floating canopy spreading a generous shadow, like the roof of a Francis Kéré building, or a sophisticated system of fiberglass fabric that opens and closes with the path of the sun, like the dynamic façade of the Al Bahr Towers in Abu Dhabi. Or a canopy of photovoltaic panels that generate clean energy as they block sunlight, like the agrivoltaics on Greg Barron-Gafford's research plot, or the dramatic power parasol that covers a pedestrian plaza on the campus of Arizona State University (ASU) in Tempe.

Whatever you're imagining, it probably doesn't resemble the shade structure that the Los Angeles Department of Transportation (LADOT) unveiled at the press conference: a flimsy, two-foot-wide, maybe ten-foot-tall, skateboard-shaped pierced-metal strip, bolted to a bus pole on the side of the road. Although it was a cloudy day, it was apparent that the minuscule sunscreen could not cast more than a column of shadow for one, *maybe* two people, and there wasn't even anywhere to sit. A telephone pole protected more passengers from the sun. The triumphal tone for such a physically tiny achievement was too much.

A Streetsblog reporter snapped a few photos of the structure and shared them with the world on Twitter, and it wasn't long before disbelieving users started throwing a different kind of shade. Another photo posted by the shade's designers, Kounkuey Design Initiative, highlighted La Sombrita's second feature, a solar-powered streetlight stashed in the structure's visor-like brim. The image captured Connie Llanos, the interim director of LADOT, in what appeared to be a grimace. Kounkuey's designers had traveled to Quito, London, Vienna, and Hamburg to study gender-inclusive transportation systems, and they came up with this? The whole event, a *Bloomberg* writer noted, "seemed unintentionally engineered for internet comedy, a sketch from *Parks and Recreation* by way of *Spinal Tap*." Or *Zoolander:* What is this? A shade for ants? After countless guffaws and retweets, the photos of La Sombrita's grand reveal were seen over twelve million times on Twitter, the result of so many dunks on the wrong kind of viral tweet.

Real-life Angelenos weren't having it, either. A reporter from Univision caught wind of the furor and took to Third Street to ask bus riders what they thought. No one was impressed. "Well, it doesn't cover anything," someone said. The lamp was too dim to be useful. "This is ridiculous," an elderly passenger in Boyle Heights, a Latino neighborhood where another of the four prototypes was installed, told a *Los Angeles Times* video reporter. Why don't they at least move it near a bench, so someone can sit down? Inside city hall, the mood was dour, as Mayor Karen Bass's public works team scrambled to contain the fallout from a project they had learned about less than twenty-four hours before, in the form of a press release that, as if anticipating the derision, had no images.

Online, La Sombrita became unmoored from the facts. The city paid *how much* for that thing? Although the sunshade cost around $10,000—not bad for a prototype, and much less than L.A.'s standard-issue bus shelters—the design and fabrication were grant-funded by Kounkuey. No city money was used. It became an icon of municipal dysfunction, all the same. "If this piece of metal doesn't represent government bureaucracy and incompetence, I don't know what does," a TikToker chided. "Another example of the weird paradox of American modernity: a society so rich it cannot afford to do anything," sneered the tweeter behind the popular Politics & Education account. "But because we usually can't do anything at all, this must be praiseworthy." Here was what appeared to be another hapless city government taken for a ride by a bunch of consultants masquerading under the banner of DEI. "Performative grift," someone else sneered. "It looks like a scam with upbeat PowerPoints," snickered another tweeter. As I sailed through the tweets, and later stood in Sombrita's pitiful shadow, I had one question on my mind. What was it going to take for L.A. to finally make some real shade?

Ask anyone who doesn't live in L.A. and they won't hesitate to tell you that it's a city of drivers. They've seen *La La Land* and *Drive* and *L.A. Story,* movies where characters are always in their cars and

A woman waits in the slim shadow of La Sombrita, an urban shade structure in Los Angeles.

stuck in bumper-to-bumper traffic. They've seen that iconic footage of the police chasing O. J. Simpson in the Bronco on the freeway and they've played *Grand Theft Auto*. And they've seen *Speed* and watched *Insecure,* where characters played by Sandra Bullock and Issa Rae are reluctantly forced to ride the bus because they've lost their license and crashed their car, respectively. But as the Pacoima community organizer Felipe Escobar reminded me, there are lots of Angelenos who ride the bus not as punishment, but as a matter of course. On the day that La Sombrita was unveiled, around 707,000 passengers boarded well over two thousand vehicles covering around 120 routes through most neighborhoods and cities in the county. Most of the bus riders weren't white-collar commuters trying to avoid a parking headache, but rather, impoverished Latinos who cannot afford a car, and rely on the service to traverse a sprawling urban desert.

And like lizards of the Southwest, they were made to hunker down in whatever shade they could contrive. Despite the high ridership numbers, government investment in bus service tends to pale in comparison to other forms of transit, like trains. Even though the bus accounts for more than three out of every eight rides that Americans take on public transportation, major transit agencies spend only 6 percent of their station budgets on shelters, signs, and benches at bus stops. Which is to say, the fact that there's hardly ever purpose-built shade at the bus stop isn't a uniquely Los Angeles problem. What is unique to L.A. is the sheer number of people who usually end up waiting in the sun: easily three times as many daily passengers as Houston, four times as many as Las Vegas, and ten times the number of riders in Phoenix, where three-quarters of the city's four thousand or so bus stops have a dedicated shelter.

The architect Doug Suisman once compared L.A. bus stops to medieval pillories, places where social pariahs were punished by way of "humiliating public confinement." The amenities, such as they are, consist of an ad-festooned bench with anti-homeless armrests and a route number sign. Passengers sit inches from the curb, with no protection from the vehicles speeding by, nor shelter from the sun or rain. Refuse overflows from nearby trash cans. "The punishment is completed by hundreds of scornful glances from passing drivers, comfortable in their air-conditioned cars. An Elizabethan constable would have heartily approved," Suisman wrote. The conditions might have been tolerable if the bus came regularly, but in L.A., about half the journey time of a typical ride is actually the wait. And that doesn't include the walk there.

La Sombrita was supposed to be one way to improve those abysmal conditions. It was intended to throw more shade on scorching sidewalks where thirty minutes of sun could cause an *abuela* to pass out, and to shine more light on dark streets. It was an earnest attempt at making public transportation safe, comfortable, and accessible—all things that drivers, sitting in their air-conditioned cars, take for granted. And it was trying to tackle a worthy issue of gender equity by focusing on the subtle ways that women use the bus system differently from men. Could a tiny, teal-painted hunk of cheese-grater metal really make a difference on those fronts?

Naria Kiani, a former senior planning principal at Kounkuey who oversaw La Sombrita, said the process began with data. It's not that women heat up faster than men, although some scientists think that may be true. Kiani and her colleagues surveyed or interviewed well over four hundred bus riders in three different neighborhoods to grasp the challenges. Many respondents were not commuters heading downtown, but female caregivers who were trip-chaining, which is transit-speak for getting on and off the bus to run errands on the way to an ultimate destination. Instead of sitting behind a desk, they were standing in the sun in the middle of the day. More transfers, more waits, more time on the street: It wasn't a surprise that heat and the lack of shade were major concerns. "Several women mentioned that they travel with kids and they travel with elders," Kiani told me. One interviewee said the environment was "life-threatening." Kiani didn't hear much about the bus ride itself.

How could Kounkuey help them? The firm planned to offer

LADOT a multitude of options to improve service for women, and through a grant from the Robert Wood Johnson Foundation, pilot one with a small-scale trial. But which? They knew long waits were a problem, but they couldn't change the bus routes. The designers knew women felt unsafe when they waited alone, and they considered bringing street vendors to the stops, along with their rainbow umbrellas, dangling lights, and eyes on the street. Kiani said that proposal was rejected. Instead, Kounkuey decided to physically improve the stop. But based on the way bus riders have long been treated in L.A., it was easy to interpret La Sombrita as an empty gesture at best, or an insult at worst. Angelenos have waited for shade for a long time, and this was not enough. Arriving amid such fanfare, the pitiful shadow seemed like another broken promise.

The saga begins in the 1970s, when sunshine and car exhaust conspired to make L.A.'s smog the worst in the nation. On bad days, California regulators issued smog alerts, urging drivers to take public transit for the good of the environment. It was a sensible request, but in a sharp 1975 op-ed in the *Los Angeles Times,* a Beverly Hills activist named Ellen Stern Harris said there was "no way" middle-class people like her would ever leave their cars in the garage unless they weren't made to feel punished for it. Harris wanted a "minimum of convenience and comfort" while she waited twenty-five minutes for the bus, and thought a bench and a shelter might go a long way.

Market research and academic studies have since confirmed Harris's argument. Clean, dignified, and comfortable shelters entice new riders. Shade and a seat make a bus stop, and a sidewalk more generally, accessible to the elderly and people with disabilities. They make streets feel safer for women. Riders at stops without these and other amenities say a ten-minute wait feels more like a twenty-one-minute drag, according to surveys conducted by University of Minnesota researchers. Along with other sidewalk improvements, like more signage, better lighting, and smoother pavement, shelters have increased ridership by 31 percent at stops in North Carolina and 141 percent in Utah.

But in Los Angeles, getting shelters in the ground has been surprisingly difficult. There are nearly 1,900 official bus shelters within the city limits, but their distribution has been governed by profit, not need. Instead of transit planners calling the shots and strategically placing shelters outside grocery stores and doctors' offices on high-frequency routes, the job is outsourced to advertising companies. L.A.'s first thousand bus shelters were installed in the 1980s and 1990s by Shelter Media Associates in exchange for the right to sell ads on the wall panels. It was an attractive proposition: The billboard company would install and maintain the shelters at no cost and pay the city more than $1 million in ad revenue per year. But because the program was designed to sell ads, not shade riders, the company tended to place the shelters in wealthy areas where ridership was low. A program that was supposed to improve the lives of Angelenos and encourage everyday environmentalism had turned into a meager tax replacement scheme for a cash-starved city.

In 2001, the mayor signed a deal to double the number of shelters and give public officials greater control over their placement. The new vendor, Outfront/JCDecaux, agreed to install and maintain shelters throughout the city and offset its costs with freestanding ad kiosks in lucrative areas. But the deal broke down because of an arduous approval process that required sign-off from the city council, public works department, eight city agencies, and nearby property owners to install a single shelter. Theoretically, that arrangement was democratic, because it allowed more public input than a shorter, more efficient process. In practice, it was a policy of inaction, scuttled by politically savvy constituents who didn't ride the bus and objected to the eyesore of advertising or the possibility of an unhoused person sleeping on their street. Two decades later, when the contract came to an end, the vendor had added about 660 new shelters, roughly half the projected number, leaving only a quarter of L.A. bus stops with dedicated shade.

Instead of reforming the broken shelter system, L.A. officials tried to work around it. In 2020, Christopher Hawthorne, the city's chief design officer, convened a shade structure design competition, seeking cheap, durable, and scalable solutions to the city's seemingly intractable sun problem. Kounkuey submitted drawings of large, colorful sunscreens with rear-mounted benches, and bright, circular railings installed around tree trunks, to take advantage of their leafy shade. But nothing ever came of the competition, because there was no public funding to implement the clever designs. Instead, Angelenos got a handful of green patio umbrellas, shading a scattering of metal benches around the city. Umbrellas aren't a bad idea, and in fact, since 2015, Seoul, South Korea, has installed over 3,400 cantilever umbrellas on sidewalks to mitigate summer temperatures. L.A., however, had no long-term plan for its public parasols, and as they fell into disrepair, city crews gave up on replacing them.

La Sombrita arrived at a time when there was more heat on shade than ever before. Climate change was raising temperatures, heat waves were becoming more extreme, and all the while, bus riders were still standing in the sun. Most of them could not afford to live in a cool, shady neighborhood, nor blast air-conditioning at home. Now, their advocates argued, they were more likely to suffer heatstroke, simply because they were too poor to drive. Didn't they deserve the same protection from the heat as their wealthier, whiter counterparts?

"Because we don't have enough shelters, because we don't have enough benches, because we don't have enough amenities, they suffer there, and they suffer more because of the worsening climate crisis," concluded Mike Bonin, a progressive Westside city councillor. During a record-breaking September 2022 heat wave, Bonin took the bus to see the brutal conditions for himself. The sights appeared to radicalize him: People pressed against a building to avoid the sun, "as if they're on the edge of a ledge." An elderly man, about to pass out, trembling as he accepted a water bottle. Given the demographics of the city's bus riders, "not only is this a public health issue, it's a civil rights issue." And then, in 2023, came La Sombrita. If this is what city officials considered success, then clearly, Angelenos were in trouble.

Spending Political Capital for the Public Good in Barcelona

As the Twitter dunking continued, Kiani was on the phone in her Echo Park backyard, patiently explaining how L.A.'s latest spatially constrained shade structure came to be. It wasn't just the internet that thought it was too small. The designers did, too. Initially, La Sombrita was larger, with an overhead canopy and a bench. One version boasted multiple connected screens, like a room divider between the sidewalk and the sun. "We started out with this very ambitious plan of what the shade device could be," Kiani said. "And as we understood the regulations more, and had more conversations with city departments, we found a path to something that could work today."

A more substantial structure could have triggered the same laborious permitting process that scuttled so many bus shelters before. Moreover, it could have required reengineering the sidewalk. To avoid those outcomes, Kounkuey instead redesigned the piece of furniture that had already been permitted: the bus stop sign. This meant their shade structure had to be much smaller and lighter. The overhead canopy shrank to a brim that did not protrude into traffic nor the passage on the sidewalk. The screen became perforated, so the wind could not blow the pole out of the ground. And the seats disappeared, to avoid installing legs that would touch the sidewalk and enter a second agency's purview. Ultimately, La Sombrita's shadow was not shaped by efficacy. It was whittled away by red tape.

Kiani reminded me that La Sombrita was a pilot; the plan was to get something in the ground as fast as possible, collect feedback from the public, and reevaluate. If transit users demanded a bigger, more substantial shade and maybe even some seats, that's what Kounkuey would advocate for in the next iteration. But at the same time Kiani knew that in some neighborhoods La Sombrita would be the best they could do. "This is supposed to be a solution for narrow sidewalks, where a bus shelter wouldn't work, that would get no solution," she said.

The problem is that *most* L.A. sidewalks are narrow sidewalks. According to a UCLA analysis, the median width of a commercial sidewalk in the city is about seven feet, eight inches—but bus shelters require at least eight feet of space, both for the structure and to preserve four feet of clearance for a publicly accessible path. L.A. sidewalks are not accidentally small. They are narrow by design. Since the 1920s, city engineers have gradually expanded the roadways and shaved away sidewalks to ease traffic, in the process sacrificing room not just for shelters, but also trees, benches, and all the other hallmarks of pleasant and sociable streets. In fact, since the 1960s, L.A. has required developers to widen roads when they build apartments and commercial buildings, also to reduce bottlenecks. Only in 2023 did climate-minded city councillors finally ask the public works department to halt the practice. It may seem improbable that a city with a hundred years of fealty to drivers could execute a grand scheme to make its streets more livable for everyone. Yet there are leaders elsewhere who understand that pedestrians don't need *sombrita,* but dignity, shelter, and *sombra.*

L.A. officials would do well to look to Barcelona, Spain—a city with one-third of L.A.'s budget and ten times the chutzpah. Barcelona once had its own problems with cars gobbling up public space that could have otherwise been shaded. But instead of working around the problem as L.A. has, Barcelona has gradually been working to confront it. The sustained campaign to get Catalans out of their cars has culminated in a spectacular urban transformation. Since the 1980s, Barcelona has proactively reclaimed space from drivers and turned it over to walkers, bikers, buses, and trams. City officials have replaced traffic lanes and parking spots with trees and vegetation to mitigate rising heat. What made this possible was not a single pilot program but a municipal government that was reoriented to use public space and transportation networks to fight climate change.

Just like Angelenos, Barcelonans have been waiting a long time for shade. But in contrast to L.A., their city was planned to be green. In 1855, as the city began to surge beyond the confines of its medieval walls, a civil engineer named Ildefons Cerdà laid out a new district that would become today's Eixample. There, every city block would center around an interior courtyard, which would serve as a communal garden, and the main arteries would be as wide as Parisian boulevards and planted with plane trees every twenty-five feet. Unfortunately, the city's developers did not share an interest in preserving common space, and over time paved over the potentially sylvan landscapes. The small patios spared inside the blocks were not lush and verdant but cramped and gloomy, and some became parking lots. Generalissimo Francisco Franco hammered the nail in the coffin by designating Barcelona as the production hub for his beloved SEAT 600 sedan, and turned what could have been havens of green into urban highways.

After the dictatorship, the long-suffering Barcelonans struck back, urging their newly democratic city council to revive the public space that Franco's government ceded to cars. In came Oriol Bohigas, a beloved architect who, as the city's chief urban planner, began masterminding Barcelona into a new urban landscape for pedestrians, the linchpin of a successful strategy to land an Olympic bid. Bohigas transformed derelict squares and obsolete industrial grounds into elegant parks and plazas festooned with sculpture. The city reclaimed and restored Cerdà's courtyards in the Eixample. Bohigas also initiated efforts to bury the freeways and build parks and urban promenades on top. One such road, the Ronda de Dalt, now tunnels under a linear park shaded by one of Barcelona's several municipal photovoltaic pergolas. Bohigas, in other words, began working toward the urban quality that would become Barcelona's international calling card by the 1990s, its walkability.

In 2015, Barcelona's voters elected Ada Colau to the mayoral office. Colau was known as a housing activist who fought evictions, but she was also motivated to act on climate. She rolled out numerous familiar policies to reduce carbon emissions, like shifting the city's power supplies to renewable sources, but also did something no American mayor would ever dream of. She declared a war on cars. At around six thousand vehicles in use per square kilometer, Barcelona had Europe's highest traffic rate, and residents were choking on their deadly fumes. Colau aimed to take more of them off the roads, and lessen the local and global impacts of their pollution.

Roads comprise about 60 percent of Barcelona's public space, but decades of investments in sidewalks, subways, trains, and bike lanes have reduced residents' auto dependency. By the time Colau took office, cars only accounted for about 20 percent of travel within city limits. The mismatch between the space devoted to cars and their actual use presented an opportunity, and Colau ordered her administration to find new uses for the redundant space. Through a major program of traffic calming, she began stripping the travel lanes from major roads like the Avinguda Meridiana and from the main streets of outlying neighborhoods like Sant Andreu and rededicated them as tree-shaded footpaths and bike routes.

The urban designer Bohigas had an untested idea for something called a "superblock," a small grid in the middle of the city, nine square blocks in all, that dramatically reduced car access to make a large, pleasant space for neighbors to chitchat. A plan for a citywide network of these traffic-free urban islands was developed in the 2000s, and Colau's predecessor, Xavier Trias, approved an ambitious plan to build some 503 superblocks in all, adding to Bohigas's goals of social cohesion more ecological ones of cleaner air, fewer emissions, and more urban greenery. The roadblock was not planning, but political backbone, as Trias appeared to have slow-rolled the implementation. When Colau arrived, the plan took off. She implemented the first of her superblocks in the Poblenou neighborhood. After cinching traffic to a single lane, planners painted the remaining asphalt in creamy blue and funky yellow, and filled it in with benches, picnic tables, and potted plants. The changes weren't without controversy, but residents in the parks-poor neighborhood came to embrace the transformation.

To execute these ambitious plans, Colau had to reform the municipal government. The administration combined the environment, transportation, and urban planning agencies into one unified department, which reported to the city's chief architect to implement the superblocks. Instead of clashing over competing remits, as L.A.'s city departments sometimes do, Barcelona's were working together. As urban planners stripped lanes from vehicle travel, the environment department moved to fill them in with trees and vegetation. As car traffic slowed down, the transportation department ramped up service on buses. In 2016, the newly formed Ecology, Urban Planning, and Mobility Department made its ambitions known to the public: *Omplim de vida els carrers!* Let's fill the streets with life!

In 2020, the Covid-19 pandemic hit, and the slogan acquired a literal meaning. Colau had built seven superblocks when Barcelona forbade indoor gatherings to stop the spread of the virus. Suddenly, her car-light streets, which offered room for social distancing, were some of the only safe places to congregate. Public space and public health became one and the same. Colau's prescience was vindicated, and emboldened by her success, she took the next steps of the superblock program to permanently green the streets. The administration would start in the Eixample, the district that had been planned but never fully executed as a green grid more than 160 years ago.

When I visited in 2022, Colau had broken ground on her first "green axis," a twenty-one-block stretch of the Carrer del Consell de Cent, a four-lane road that was being converted to a linear park. As I walked the avenue, the concrete sidewalks and black asphalt dissolved into a single gray path bereft of curbs. The traffic faded and the parked cars gave way to café tables, fragrant shrubs, and plane trees arching overhead. Now and then, a delivery truck ambled by, but more numerous were the children zooming along on scooters. At an intersection, a single sedan ambled around an oversize traffic circle where youngsters and elders were serenely resting on benches in the feathery shade of a jacaranda.

As the first step of a larger, citywide transformation, Colau aspired to naturalize one out of every three streets in the Eixample in this fashion, ensconcing the district's 260,000-plus inhabitants in a new kind of urban green. With fewer cars and combustion engines, there could be fewer greenhouse gas emissions and less locally generated heat. By resurfacing the asphalt with bright concrete tiles and permeable granite pavers, the roads could both reflect solar radiation and evaporate its heat. And with more shade, the streets could be cooler and more comfortable. City documents call for extravagantly large beds of grassy soils, giving tree roots and their canopies more room to breathe. At the height of summer, no less than 80 percent of a green street could be shaded by their leaves. This vision is still far from reality, but Barcelona has made important strides.

Unlike in L.A., where the care of the urban forest relies on neighborly goodwill, Barcelona has achieved an ultraefficient system to nurture its street trees. Pumps and cisterns suck water from tanks and aquifers dotted under the city and shunt it through miles of underground pipes and subsurface drip irrigation tubes. The city's six hundred or so gardeners used to water the urban forest by hand, driving around the city to sprinkle saplings from the seat of a tanker truck, in the process both polluting the air and wasting precious water. That changed after the droughts of the 1990s, said Gabino Carballo, a landscape architect in Barcelona's parks department. Now, gardeners can deliver just the right amount of water without leaving their desks. Although the advantages of computer-controlled, automatic irrigation are many, above all it gives trees a better shot at survival as water becomes scarcer and urban temperatures continue to rise. No doubt, it ain't cheap. But if Los Angeles, or any other American city, is going to get serious about shade the way Barcelona has, then it may have to treat it the way Barcelona does, as an essential urban service supported by specially designed infrastructure. Carballo explained that the irrigation system expands as city streets are reconstructed, and anticipated that every tree in Barcelona will be connected by 2070.

Cooler streets are safer for everyone, even those who prefer to drive. According to a 2023 study in *The Lancet,* Barcelona's urban heat island effect causes more than 360 premature deaths every summer, resulting in one of Europe's highest such rates. Global climate change will raise the city's artificially elevated temperatures even higher. Green streets can be a solution. A citywide network could bring Barcelona close to an overall goal of shading about 30 percent of the entire city under trees. The *Lancet* researchers calculate that an urban forest of that magnitude could drop the entire city's ambient temperature by 1 or 2 degrees. By additionally removing cars, the total cooling effect could be closer to 3 degrees. In combination with other changes, more trees and fewer cars could have significant effects on mortalities and morbidities, potentially saving Barcelona billions in healthcare costs. This is no small matter in a country where the public foots the bill for universal care.

But because trees take time to grow in, Catalans won't benefit from those health effects for years. In the meantime, all the construction proved to be a nuisance, and Colau paid a political price for it. She was voted out of office in 2023 as part of Spain's right-wing lurch. While the fate of her project is unclear, her vision continues to inspire other world leaders. The United Nations Environment Programme recommends copying Barcelona's superblocks in other global cities that will have to adapt to heat. European architects, planners, and city officials continue to flock to Barcelona to see them for themselves. No city has embraced the green street movement like Paris, where Mayor Anne Hidalgo is removing as many as 60,000 parking spaces and filling them with verdure. Amsterdam, Brussels, Vienna, and Copenhagen are also reclaiming car habitat for trees. Even Los Angeles might take a crack at a car-light green street in the form of a "park block." Perhaps it has learned from the debacle of La Sombrita that shade should not be wedged into a street. Instead, a street should be redesigned to accommodate more shade.

Shade as Critical Infrastructure in Singapore

For all its faults, there are some things that La Sombrita got right. One, it made shade. Two, it made shade in the right place. It is seldom appreciated by city officials that at L.A.'s latitude, the sun is never exactly overhead, even at the summer solstice, and for that reason, shadows are never cast directly on top of us. At perfectly east- and west-facing bus stops, shadows from bus shelters at the curb are often cast in the road. And at south-facing stops, the shade is not usually found under a shelter roof, but behind it, fanning across the sidewalk. In a city where it's warm all year, it makes some sense to locate the waiting area behind a shelter, too—or perhaps behind a vertical screen, perforated for visibility. La Sombrita's designers did just that.

They made sure the sunshade's little shadow cloaked the passengers on the sidewalk, not the cars in the street.

"I started doomscrolling through all these tweets, and I was like, man!" said Ariane Middel, an urban climate scientist who runs ASU's Sensable Heatscapes and Digital Environments (SHaDE) Lab. "What is wrong with all you people?" For about a decade, Middel has dragged a multidirectional radiometer mounted on a garden cart through streets, schools, and parks around Los Angeles and Phoenix, recording microclimate data. Middel has an innovative way of measuring shade's impact. In addition to ambient temperature, she calculates the amount of radiative heat in a particular space gained from the sun and the surrounding warm surfaces using an index called the mean radiant temperature. Middel has used MRT to determine that objects with no effect on air temperature, like umbrellas and plastic sails, are still effective personal coolers, so long as they are not located so close to the body. In fact, in arid and overdeveloped environments, she recommends them as perfectly acceptable substitutes for trees.

According to Middel, on dry and sunny days, the direct receipt of solar radiation is the main factor in human heat load. She hadn't examined La Sombrita in person, but based on her observations of other smoldering sidewalks, she estimated that someone standing in its shadow would feel 20 to 30 degrees cooler than someone in the sun's glare a few inches away. Of course, La Sombrita couldn't replace a proper bus shelter, "but that's not what they're meant to do," she continued. Los Angeles has a larger problem: the existence of what Middel, Jennifer Vanos, and Kelly Turner call "shade deserts," urban environments where there's hardly any outdoor cooling at all. The four hardscrabble neighborhoods where La Sombrita pilots were installed are bereft of the sun protections that the three heat experts say are critical to preserve human health. In shade deserts, it's common to walk twenty minutes down a barren street and wait another twenty minutes at an unsheltered bus stop to take your grandchild to a playground without any trees or shade sails. The absence of shade in these communities is not unlike the absence of nutritious groceries in food deserts. In both cases, the absence of a critical resource exacerbates health disparities.

Authorities usually advise people to seek shade in emergencies, but by the time someone's body temperature reaches a dangerous level, shade is not enough. Once heat illness takes hold, first aid is needed for safety. Middel, Vanos, and Turner believe shade is a preventive measure, not a cure, and must be ever present to avoid the prolonged sun exposure that leads to acute situations. This kind of coverage cannot be achieved with La Sombrita's two-foot puddle of shade. The eradication of shade deserts requires a more holistic, multifaceted approach, and to achieve that, they believe L.A. needs a lot more shade—trees, canopies, and buildings oriented to cast shadows, all things that they classify as "shade infrastructure."

The city that might have the best shade infrastructure of all is Singapore, the sweltering island nation. People here have long had their own tricks to deal with the torrential rain and sticky heat. Chief among them might be the covered sidewalks. Like the toldos of Seville, the origin of this public shade is also unclear. Although these "five-foot-ways," which tunnel through the ground floors of arcaded shophouses, resemble the porticoes of Bologna, they may be native to Southeast Asia. Stamford Raffles, the British colonial official considered to have founded Singapore in the early nineteenth century, wrote them into the first town plan in 1822. Raffles mandated clear, continuous, and covered passages on both sides of every street to ensure efficient transit in inclement weather.

Over time, his "verandahways" fell out of favor. They were revived in modern form by Lee Kuan Yew, the powerful prime minister who guided Singapore to independence in the 1960s. Lee was something of a micromanager and had a particular interest in climate and comfort. He believed that humidity was stifling the country's economic productivity. Indoors, he transformed Singapore into what journalist Cherian George called the "air-conditioned nation." Outdoors, he was fanatic about shade. Lee was known to lecture subordinates about the poor design of footpaths and promenades, sometimes kneeling on the burning hot ground to prove a point.

In the 1960s and 1970s, as Lee's authoritarian government erected towering public housing estates, architects kept the ground floors of every building open to the air, preserving the areas as communal "void decks" where residents could gather to catch a breeze. In the 1980s and 1990s, Singapore's housing and transportation agencies directed the construction of freestanding metal canopies over the sidewalks to ensure dry paths from the estates to the nearest bus or train. Today, the authorities claim to have erected around 125 miles of covered walkways. Try to imagine if New York's ubiquitous construction scaffolds were permanent sidewalk architecture and you might have some idea of what the immensely unattractive though functional achievement looks like. In America, real estate developers are required to set their buildings back from the street to let in more light, but in Singapore, they must contribute to the shade network by carving eight to twelve feet of pedestrian overhangs out of the ground floors of their buildings.

Research suggests the canopies have an effect similar to that of a clean and well-designed bus shelter. Just as a shelter can make a wait for the bus go by faster, so too do Singaporeans report that a stroll under the walkways feels 14 percent shorter than a stroll under the sun. (It helps that Singapore is located 1 degree above the equator, which makes the sun's annual path more stable and the earthbound shading simpler to plan.) When it comes to shade, most everyone would prefer the leaves of a luscious canopy to a clunky aluminum roof, but trees can't always be the answer, said Lea Ruefenacht, a former researcher with Cooling Singapore, a government-affiliated urban heat initiative. She reminded me that trees create cooling through shade and by releasing water into the air. In humid Singapore, more moisture can add to the misery. For comfort, Ruefenacht recommends a balance of both green and gray shade.

In Singapore, the densest gray shade is found in the concrete understory of the skyscraper forest downtown. Real estate developers are required to furnish what the authorities consider "sufficient" shade on outdoor plazas, cooling at least 50 percent of seating areas between 9:00 A.M. and 4:00 P.M. The shade can come from any number of sources—trees, umbrellas, awnings—but in their design circulars, the authorities demonstrate that it can also be afforded by a nearby tower's knifing shadow. This approach can be contrasted to that of New York City, where building shadows on outdoor spaces are discouraged, and the mere threat of their existence can scuttle a new development. In this cooler climate, developers are instructed to site their plazas on sun-facing south sides, to create winter warmth. (In fact, the plazas are not allowed to face north.) Singapore has a different approach. Ideally, developers locate plazas on the east side of their buildings, so they can be cooled by afternoon shade.

It is the rare place where urban shadows are encouraged as a public benefit. "In the tropical regions of the world, part of the problem has always been that settlements inherit building codes from the temperate regions, and they don't necessarily have the means to review it and ask, does this work for us?" explains Kelvin Ang, the conservation director at Singapore's Urban Redevelopment Authority. "In Singapore, somehow there was a lot of awareness that building codes and planning codes had to encourage shade, because of the intensity of the sun." Planners believe that if a public space is unshaded, no one will use it.

Despite the potential effects on humidity, Prime Minister Lee demanded trees everywhere, believing that a "clean and green" Singapore would be attractive to foreign investors. Under his command, a newly formed parks and trees unit spruced up the major boulevards, embowering them under the broad canopies of Angsanas, rain trees, mahoganies, and acacias. "Flowers are okay," Lee reportedly told the department head, "but give me shade first." In the 1970s, as he implemented congestion pricing and other schemes to push Singaporeans out of their cars and onto public transit, Lee turned his attention to the sidewalks, crosswalks, and bus stops where a pounding sun could have repelled potential new riders.

In Los Angeles, trees are the last piece of the street design puzzle, blasted into concrete pits and stuffed haphazardly into sidewalks after every vault and meter has been trenched, every curb has been built, every gutter cut, and every driveway poured. In Singapore, however, Lee ordered his land use planners to consider them from the beginning. The overhead power lines that disfigure L.A.'s sidewalks and make trees small and shrubby are rare. Most utilities are laid underground in vaults that run alongside the street trees and their roots. The green infrastructure is plotted by urban planners, engineered by public works agencies, and managed by a parks board whose budget increased tenfold under Lee's leadership. The funding and coordination have proven to be the difference between a thriving urban forest and a bunch of sad city trees.

Besides the roads, Lee's urban planners mandated greenery in private developments, regenerating a new garden city to compensate for the natural rainforest that was all but gone. The Singaporean government had a lot of leverage. Through strong eminent domain rules, it owned about 90 percent of the land, and building inspectors wouldn't clear a building for occupancy until they saw trees in the ground. Singapore's extensive public housing estates also came with grassy lawns, leafy courtyards, and tree-lined paths that connected to parks and nature preserves. As a result, trees are just about everywhere in Singapore, in rich and poor neighborhoods alike. "We did not differentiate between middle-class and working-class areas," Lee wrote in his memoirs, claiming it would have been "politically disastrous" for the People's Action Party. This makes Singapore distinct from American cities, where shade is a reliable indicator of economic inequality.

Thanks to Lee's smart planning policies, including developing thousands of acres of local parks and hugely ambitious land reclamation efforts, Singapore managed to do something remarkable: It became simultaneously denser and greener. The authorities claim the urban forest grew from 158,600 trees in 1974 to 1,400,000 trees in 2014, even as the city added three million more people. Today, more than half the island is covered in grasses, shrubs, and broad-canopied trees, throwing cold water on the idea that cities can't spare room for nature as they grow. "It's the biophysical environment that's a differentiating factor," said Daniel Burcham, a former researcher at the parks board, when asked to explain Singapore's success. "It's just easy to grow trees when it's summer every single day and you have over two meters of rain every year." But without political consensus, there would not have been room spared for those trees to grow. "This was a goal that they were going to pursue, and it was a vision that they were all united around achieving."

Burcham now teaches arboriculture at Colorado State University, in Fort Collins, a semi-arid city where political leaders have a few years in office, not decades. "Some would characterize Lee Kuan Yew as a strongman, or semi-authoritarian figure, and to some extent, that's very true. But this is one good thing that came from that system. He set out this goal, and provided material resources and provided political support for people to achieve it," Burcham explained. Although it would require coherence across administrations, there's no reason in principle why democratically elected governments in tropical cities like Miami or Honolulu could not also sustain such a project.

Does all this shade protect Singaporeans? In the afternoon, the streets of Singapore's business district, plunged in the shadows of skyscrapers, are the coolest in the city. The effect ends when the sun goes down, and the buildings release the solar radiation they absorbed. At night, the green grounds of a public housing estate may offer the most relief, as the air is 2 to 4 degrees cooler than the drafts whistling through a bustling commercial strip. The well-established epidemiological link between air temperature and heat illness would indicate that these shadiest neighborhoods are indeed Singapore's safest from heat. Shade infrastructure like trees and buildings won't be enough to overcome all the warming effects of climate change, but it will make a difference.

It's unlikely that local American governments can be as effective as Singapore's, an autocratic nation-state long ruled by a strongman with a personal interest in shade. Nor are most American cities fortunate enough to have Singapore's ideal climate for growing trees. Nevertheless, Singapore shows what can be done with intentional government planning of shade. A cooler city for everyone is within reach. Let's not pretend it's impossible.

Australia's Shady Public Health Campaign

What if L.A. residents actually liked La Sombrita? What if a year's worth of feedback from bus riders in Westlake, Boyle Heights, Watts, and Van Nuys showed that some shade was better than none at all and that the night-light was a comfort on dark streets? The transportation department would be delighted by the results. Instead of committing to a billion-dollar project to rebuild and expand thousands of miles of crummy sidewalks across the city, they could instead drill in Sombritas. But even if every bus stop has a Sombrita, it would comfort only one or two people waiting for the bus, and not the eight or ten people standing beside them.

If L.A. wants to tackle shade in the most advantageous way, then what it needs to do is think about shade provision not at the scale of the individual—as a form of personal comfort—but as a public health solution at the scale of the population. To begin with, the leaders in L.A. and other American cities could look to Australia, a country that has its own population-level problem with sun exposure and is meaningfully addressing it for the good of everyone.

In Australia, the threat has not been the sun's heat, but its powerful ultraviolet (UV) radiation, a form of energy that destroys skin tissues and causes them to regenerate as cancers, especially in white people. The situation is dire, with about two in three Australians predicted to receive skin cancer diagnoses during their lifetime. Melanomas and carcinomas cost the country around two thousand lives and more than a billion Australian dollars in healthcare treatments every year. The good news is that a sustained, society-wide fight against skin cancer is succeeding. Although Australia has a long way to go, incidence rates are falling among younger generations that have been targeted by the initiative. The United States could follow the country's lead and implement some of the same strategies to get ahead of the curve on the other effects of sun exposure, like the heat deaths and illnesses that are already rising.

Like many Americans today, Australians worship the sun. Australia is the "sunburnt country," as the beloved Dorothea Mackellar once wrote. It's long been a point of pride for the fair-skinned blokes and sheilas to bronze their bums on the beach. Although epidemiologists had been tracking rising melanoma rates since the 1950s, it wasn't until the late 1970s and early 1980s that the sun came to be a cause of concern. Not long after, hysteria mounted over a thinning layer of ozone over the South Pole that was believed to be exposing Australians to more UV. At the time, around thirty out of every one hundred thousand Australians were diagnosed every year with melanoma. Today, that rate has more than doubled.

Public sentiment began to change, as it sometimes can, with a jingle. In 1981, a friendly, lisping cartoon bird named Sid the Seagull began dancing across Australians' tellies, urging kiddies to "slip on a shirt, slop on sunscreen, and slap on a hat" to prevent skin cancer. "Slip! Slop! Slap!" was Australia's own "stop, drop, and roll," a mnemonic seared into the national consciousness through TV ads, PSAs, posters, and eventually a school curriculum. Although a youngster's sunburn significantly raises the risk of melanoma two decades later, the Cancer Council of Victoria, the regional nonprofit that commissioned the tune, decided to target the awareness campaign to children because they are more impressionable than adults. "Habitual behavior starts in childhood," recalled one of the campaign's founders. "We'd park at the beaches with the jingle playing and give out stickers and free tubes of sunscreen. Kids and families would flock to us and sing along. It was amazing to see how many people knew the song."

The national Cancer Council pushed for free sunscreen dispensers at public beaches and encouraged parents to dress their children in UV-protective shirts and hats. Adults who worked outside could also claim tax exemptions for sunproof gear. Next, the regional councils began certifying SunSmart schools that adopted sun protection policies. Certified daycares and elementary schools forbade children from taking recess if they did not have a wide-brimmed hat. ("No hat, no play" is the other mnemonic drummed into children from an early age.) Gradually, Australian social norms began to shift away from suntanning and toward sun protection. What started as a media campaign to change individual behavior evolved into a more coordinated effort to change the communal behavior of schoolchildren. Then came the campaign for environmental change. It wasn't long before Sid the Seagull urged Australians to "pick up a spade and plant some shade."

As in the United States, the presence of shade in urban parks and schoolyards is an indicator of economic inequality in Australia. A study in metropolitan Sydney found that shade was more robust and concentrated in playgrounds in wealthier neighborhoods. The Cancer Council believed that a strong campaign to raise awareness of shade's benefits could cause more schools and local governments to invest in its provision. "Cancer organizations around the world tend to believe a lot in messaging, and messaging of course is important," said John Greenwood, a Sydney-area architect. "But my angle is, if we message people and tell them to get in the shade, somebody has to make sure that there's an effective piece of shade for them to get into." In the 1990s and early 2000s, Greenwood was that somebody. As a shade consultant, he assessed Sydney's Olympic Park and numerous other grounds and fields, recommending tensile membrane roofs, vine-covered pergolas, and sun-blocking glass over picnic areas, public bathrooms, and "anywhere people are queueing and gathering," he told me.

In 1998, New South Wales's Cancer Council and health department commissioned him to write *Under Cover,* a comprehensive, 190-page shade-planning document for parks, schools, playgrounds, swimming pools, and other outdoor sites where sunburn was a risk. In addition to a lecture tour, Greenwood estimates that he conducted somewhere around 250 "shade audits" at sites across Australia, New Zealand, and Canada, inspections of all the things at a particular location that cast shadows, or what Kelly Turner would call the shade infrastructure. As in Singapore, this included not only trees and canopies but also buildings.

One goal of a shade audit is to help a client make better use of the resource that already exists. At a Sydney elementary school, Greenwood charted the seasonal shadows cast by the school buildings, the covered walkways, an outdoor shelter, and a row of eucalyptus trees along the perimeter. Although the structures offered what he deemed to be good shade, he noticed the gum tree shadows were inaccessible in the summer because they were cast in the garden beds. Greenwood suggested the school prune the trees' lower branches and relocate some seats and tables from unshaded locations to the area under the canopies. Then, to create the additional 2,600 square feet of cover that he believed the school's 330 students required at lunchtime, which they all took outdoors, Greenwood recommended a row of poles to span fabric cloths across an unused footpath. Parents could demount the sails in winter, when the risk of sunburn is lower, as volunteer shade stewards. The total price tag would be a relatively modest 8,000 to 10,000 Australian dollars.

Greenwood does not eschew expedients like shade sails. They work well over sandboxes. The fabric can block more direct UV than all but the densest tree crowns, and compared to tree roots, the structural footings require hardly any room underground. Sails may be the most effective form of sun protection for adolescents who bristle at mandatory hat and sunscreen rules, according to an observational study of more than fifty high schools conducted by researchers affiliated with the Cancer Council Victoria. And most importantly, sails are effective immediately, if the supports are planned and located correctly. They do not need years to mature. But Greenwood doesn't usually recommend them as long-term solutions, because canvas, polyethylene, and polyester fabrics break down after five years. Nor do the sails offer much protection against the indirect UV reflected by the surrounding surfaces. Sails protect people, not places. Although a schoolyard of trees, grasses, and ground covers requires more maintenance, it also removes more solar radiation, both infrared and UV, than a schoolyard paved under concrete. And it's cooler. Ultimately, Greenwood recommends a combination of both "natural and built shade." Trees and sails can work together: If all goes well, by the time the fabrics fall apart, the canopies will have grown in.

Greenwood has made his software available to the public so that schools and parks can run their own shade audits, but for now his level of sophistication has proved technically challenging to all but the experts. Nonetheless, thanks to people like him and the awareness campaigns waged by the Cancer Council, Australia leads the world in protecting children from the sun. The regional chapters run their own modified audits, hand out guidelines to school administrators, and train interested parents to assess their school's shade infrastructures. Shade is now ubiquitous in Australian playspaces: A survey of thousands of New South Wales public playgrounds found that 81 percent were shaded by trees or sails, and a smaller survey of primary schools in Victoria showed similar coverage. By comparison, only 33 percent of American playgrounds are shaded at solar noon, according to a survey of more than 100 parks and schools across the country.

The shade appears to be having a positive effect. Population surveys in Melbourne reveal that people of all ages are more likely to cover up or stay out of the sun than they were in the late 1990s. Fewer of them have positive associations with suntans. Although the incidence of melanoma continues to rise among older Australians, the rate has fallen among people under thirty years old. In combination with other changes, such as banning tanning salons, the tide may be turning on skin cancer. "It's kind of like saying we've had a success story on tobacco in this country," said David Buller, a scientist at Klein Buendel, a Colorado-based public health research firm. Although many Americans still smoke, overall rates have declined with each passing generation. Buller believes Australians are on a similar trajectory with sun exposure. In the 2010s, Buller and his colleagues installed more than 140 UV-blocking shade sails over benches, picnic tables, and bleachers near playgrounds in Melbourne and Denver and compared the impacts. Overall, the use of shaded spaces nearly quadrupled, but the boost was stronger in Denver. Buller thinks it's because there are already plenty of shady playgrounds in Melbourne.

Although ozone depletion does allow more UV to strike the earth, closing the ozone hole may not have improved Australians' health all that much. The real threat was the ignorance of fair-skinned white people who did not appreciate the danger of the sun. The notion that Australians would ever be healthy by hiding behind skateboard-shaped screens at the bus stop is patently absurd. Instead, as a society, they have been saturated with messages about the importance of sun protection and have come to believe that everyone should be able to access shade. True, not all towns in Australia are shady nirvanas. Many Australian suburbs are as sun-blasted as their American counterparts, and their wide roads put some of ours to shame. But even in environments where shade is absent, Australians do not dismiss it as an unnecessary luxury. They understand that shade is a requirement for public health.

Bringing Global Ideas Home to the United States

What L.A. has done about shade hasn't worked. It has led to tweaks and so-called innovations that are incommensurate with the need. La Sombrita is small, because street space has been largely turned over to cars and other urban infrastructure. It's lightweight, because city agencies are not directed to work together to offer more durable solutions. And the four prototypes are together shadowing eight or ten riders, because the city hasn't yet invested in the provision of shade for hundreds of thousands of others. La Sombrita is the product of a city that is still thinking short term and working within the confines of twentieth-century rules to address the problems of the twenty-first. If Los Angeles wants shade for everyone, whether from a bus shelter, a tree, or any other type of shade infrastructure, then it needs to begin expanding opportunities for shade in new, holistic, long-term plans for its streets, parks, and playgrounds.

From Barcelona, L.A. should learn that shade requires political courage and even sacrifice. By deprioritizing cars, cities can use streets, their largest public space, to fight climate change and protect vulnerable people. L.A.'s political leadership can begin to pursue a shadier, healthier city by combining the public works and transportation departments into a single infrastructure agency and uniting them with a shared vision. Or the mayor could align her government behind a capital infrastructure plan, a somewhat standard planning document that cities use to guide their future investments, that prioritizes shade. The 2028 Olympics are only a few years away, and with Mayor Karen Bass pledging car-free games, now is the time to build robust pedestrian infrastructure. As Ada Colau learned, it will be a bitter pill for drivers to swallow. But if Bass is serious about urban heat, she must implement a plan that both mitigates climate change through reducing emissions and adapts to it by making more shade. Individuals make decisions based on their own short-term needs. It takes central planning to undertake long-term projects for the good of everyone, even if they are unpopular now.

You may be tempted to say that Los Angeles could never be Barcelona, just as the United States could never be Spain. But some neighborhoods near downtown L.A. are nearly as dense and transit accessible as those in Barcelona, including Westlake, where La Sombrita casts her shadow. A shade-equity district in L.A. could become a model for the rest of the country. After all, the problem of shade deserts in historically redlined communities is not limited to Los Angeles: Atlanta, Chicago, Washington, D.C., and New York are just a few American cities that are ripe for transformation. "In talking to constituents, they want more parks, more green space, but the challenge is, where do you put it?" said Hugo Soto-Martínez, an L.A. city councilman who plans to open a Barcelona-style "park block" in his district. "There are a ton of streets that could be used in this way," he told *The Washington Post.*

From Singapore, L.A. should learn to see shade as an urban lubricant. Covered sidewalks keep the city moving in inclement weather, shaded outdoor places support public gatherings, and the trees that cool the air and calm the streets make the city more walkable. Research by Anastasia Loukaitou-Sideris, a UCLA urban planning professor, finds that sun significantly impedes the mobility of older adults in Westlake who can no longer drive, even more than cracked pavement or a fear of criminals. By committing to shadier, more accessible streets, L.A. can support independent living, encourage social connections, and even combat depression among the elderly. As we all age, shady sidewalks like Singapore's will be a needed adaptation.

And from Australia, the United States should learn that shade is not just a personal comfort, but a tool to improve public health. We may be able to lessen suffering by teaching sun protection as a cultural habit and thoroughly integrating shade into our built environment. Shade alone will not protect us from rising heat, just as it does not alone ward off skin cancer. But it can dramatically improve our lives. On hot days, children feel better and have more energy to run around and exercise when they're under cover. Adolescents and adults alike are drawn to hang around shady parks. In the years to come, as heat worsens, obesity, depression, and aggression could be fueled by more time inside, hiding from the sun. Interventions like shade can help us reclaim our public terrain.

But at the moment, few American cities consider the expansion of shade a worthy goal. Only a handful include its provision in their municipal plans, according to a survey conducted by Kelly Turner and other researchers. Perhaps that's because shade's benefits are still poorly understood. Epidemiological studies of heat mortalities and morbidities generally focus on the effects of air temperature and to a lesser extent humidity, but not sun exposure. The air-cooling effects of trees are well known, but the human-cooling effects of shadows are not. It is much easier to capture the air temperature of a single representative location than it is to estimate the MRT of every sidewalk. There is no single, widely accepted index to measure shade's thermal impact.

Maybe we can learn to appreciate shade not through numbers but by analogy. In the United States, bedrock environmental laws have gradually forced the reduction of dangerous environmental hazards. The Clean Air Act has compelled state governments to scrub the chemicals that create unbreathable air. The Clean Water Act has compelled them to clear the waterways of the waste and sewage that make them unsafe. In fact, in Oregon, it even compels local regulators to mitigate solar radiation because sunlight can raise stream temperatures to dangerous levels for aquatic life. But these rules do not command such protections from the sun for our human communities. Maybe if heat was reclassified as an environmental hazard on par with airborne threats such as smog, we would think about shade differently. In such a scenario, the United States could determine that a certain amount of heat, above an agreed threshold, must be avoided to protect our health and welfare. The sunlight that warms our roads, bakes our air, and oppresses the bodies of children, seniors, and the most vulnerable members of our society would be a contributing pollutant that we would agree to abate. This would make shade a necessary remediator. In other words, if we ever agree that heat is a threat to our freedom and happiness, then we might also decide that shade, our defense against it, is an inalienable right.

So what would it actually look like if Los Angeles, against all odds, granted the right to shade?

Imagine, for instance, what Third Street might look like in 2050, after two more decades of global warming. On another 90-degree autumn morning, as you walk down the street to catch a ride on an electric bus, you might find that the route is noticeably more comfortable than it was back in the day. The car traffic is noticeably more sparse than it used to be. The once-jet-black asphalt roadway has been allowed to fade into a brighter and more reflective gray. And it has been narrowed to accommodate a row of drought-resistant rosewood and pistache trees that sway in the breeze. Dappled light filters through the leaves and the rusty screen of an abandoned sunshade that is no longer needed.

You would find yourself in new downtown-like districts where the buildings have been specifically zoned to cast shadows in September and October. In commercial areas, you might find yourself strolling a path that is cozier and more intimate than it is today, because it cuts continuously through the ground floors of the buildings. Vendors have set up shop on this covered sidewalk and they urge you to linger and have a coffee. On another hot day like this, you might order the cold brew. Later that afternoon, as the air temperature climbs, you might find the seasonable warmth more bearable without the sun pinching your skin.

In umbrageous neighborhoods, millions of Angelenos once threatened by rising temperatures would be newly shielded from heat. The risk of health problems like asthma, heart attacks, heatstrokes, and loneliness would be reduced in a shadier city. With more trees on every block, the sidewalks would be a little busier, the bus stops would be a little livelier, and on hot days the emergency rooms would be much quieter. In a city with a right to shade, someone who lives in a newly verdurous Watts could be just as comfortable outdoors as someone in the historically green enclave of Windsor Square. An elderly man walking to the store would be as safe as a wealthy lawyer walking to her Mercedes. Perhaps he is leaving his apartment, one of the remaining few without air-conditioning, to find a cooler refuge.

Do we know that all this shade is going to be enough to ensure survival on a warmer planet? No. During the extreme heat waves that transcend the physiological limits of the human body, shade will only buy us time. But these frightening events will not occur every day. The rest of the time, we will need to find a way to cope with physiologically tolerable, if still unwanted, high temperatures. Shade is one of the many adaptations we will have to make to get through unprecedented times. And as a way to improve the lives of many people in cities without making climate change worse, it is probably one of the best options we have. But on a global scale, to really protect everyone from the sun, we are going to need a bigger shade.

Chapter 9

Making Sunsets: Shade at a Planetary Scale

In January 2023, I met Luke Iseman in a Baja California Sur coastal town. The serial tech entrepreneur usually lived in the San Francisco Bay Area, but could frequently be found in Mexico cultivating a spearfishing hobby and trying to change Earth's atmosphere. With his friend Andrew Song, he had co-founded the company Make Sunsets, which purported to fire small amounts of sulfur dioxide gas into the stratosphere to reflect the sun's rays. Iseman and Song sold these vaporous plumes as "cooling credits," claiming they would reflect solar radiation and fight global warming.

On a scrubby spread of arid land, Iseman and Song worked out of an RV as local construction workers hammered two repurposed shipping containers into a more permanent home for Iseman and his business. They were preparing to launch a $159 High Altitude Science weather balloon, a favorite tool of high school science teachers, to release what would be their third-ever sulfur dioxide payload. This was the home base of the world's only active geoengineers: "a Burning Man camp with fast internet and electricity," as Iseman described it.

Iseman, shirtless and muscular, had an intense demeanor and an intimidating Mohawk. To show me how he turns cheap powder into global coolant, Iseman led me outside, where he whipped out a converted Presto pressure cooker pot and poured in about a quarter pound of Bonide sulfur powder, a chalky fungicide that he bought online. I watched as the shirtless Iseman ignited the powder with a Bic lighter and closed the lid. Inside, clumps were burning into gas that he would transfer to a white garbage bag. Eventually, the gas would be moved into the weather balloon that he would launch into the sky.

When Iseman opened the lid to inspect the process, I caught a foul whiff of rotten eggs. Sulfur dioxide is nasty stuff. It stings eyes and corrodes skin. It can cause painful breathing and even heart attacks. To avoid that fate, Iseman does his chemistry outdoors, where any smoke that escapes can blow away and become someone else's problem. Song cinched the garbage bag with two zip ties, creating a less-than-airtight seal. The two brought the bag back inside the RV to weigh the contents on a kitchen scale: 1.7 grams of sulfur dioxide, more or less. They left the bag on Iseman's yoga mat and broke for lunch.

That afternoon, they planned to squeeze the smoke along with helium into the weather balloon and drive to higher ground to launch it into the stratosphere. For this third launch, they decided to verify the balloon altitude with GPS. Thirty miles above the ground, the balloon would explode and release the contents. Iseman would capture the dramatic moment with an onboard GoPro camera. According to calculations that Iseman cribbed from the geophysical scientist David Keith, one gram of sulfur dioxide would reflect enough sun to roughly offset the amount of heat trapped by one ton of carbon dioxide over one year. Yet Iseman claimed his reflective particles, or aerosols, could stay effective for up to three years. When I asked Iseman how he reached this number, he admitted the precise details escaped him. "It's something about the wind circulation with the stratosphere that's frankly over my head," he muttered. Wind circulation was also why he released balloons from this part of Mexico. He believed particles released closer to the equator stayed in the atmosphere longer, although these particulars escaped him, too.

If Iseman is Make Sunsets' scientist, Song is the business brains of the operation. Song is somewhat less charismatic, and at a rough moment even mused to me about returning to an old job selling software. But he has lofty dreams, too. He imagined making customers out of companies that already spend millions of dollars to save rainforests. Because those forest carbon credits are mostly bunk, Song believed it was only a matter of time until green-minded businesses would be pressured to find a new way to meet their climate pledges. One day, Iseman hoped Make Sunsets would fulfill huge corporate orders by flying much larger, football field–sized balloons, or maybe a modded-out bomber jet. In a dream scenario, Coca-Cola would ask them to inject $20 million worth of sulfur dioxide above the Arctic to refreeze the ice and save the polar bears.

But those dreams were brought down to earth by government regulators and the scientific establishment. Just before I arrived in Mexico, the Ministry of Environmental and Natural Resources (SEMARNAT) caught wind of Iseman's plans and announced an intent to ban geoengineering in the country. SEMARNAT cited the many side effects that stratospheric sulfur dioxide could inflict on the Mexican people: dangerous rain and wind, extended droughts, and thinning ozone. Federal officials were evidently angered that Make Sunsets had already launched twice without bothering to ask the Mexican people for permission. Iseman was unbothered. He dismissed the decree as the work of "bitchy environmentalists."

Back in the RV, Iseman's phone was blowing up. Alex Wong, the research director at SilverLining, a Washington, D.C., nonprofit that was a leading booster of geoengineering, wanted to talk. SilverLining worked assiduously to convince the public that geoengineering, which is the deliberate, large-scale manipulation of the earth's climate to slow global warming, was not a wacky, far-fetched, or nefarious scheme but a serious subject worthy of research and federal support. And now these dudes were about to ruin everything.

Wong urged Iseman to call off future launches, but Iseman was impervious to arguments about cultural sensitivity and government oversight. "There are no regulations on oil companies," Iseman retorted. "I need to get an oil company as a client for the political clout." Wong tried a different tack. He suggested that Iseman, an enemy of the state, could now be an assassination target. "You mean I'm going to get shot by, like, *federales* over launching weather balloons?" Iseman asked, incredulously. "Yeah," came the terse reply, crackling over the speaker.

Song's theory was that Wong was spooked. The unsanctioned launches in Mexico had triggered backlash that threatened SilverLining's own planned geoengineering trials in the United States. "Obviously all they've been doing is fucking nothing," Song scoffed, and "we're doing something." After the conversation, Iseman emerged from the trailer, rattled. "This vaguely DARPA-affiliated motherfucker is like, 'You should worry about your safety if you keep launching there,'" he murmured, referring to Wong's past work with the Defense Advanced Research Projects Agency. Make Sunsets hadn't broken any laws, because there were no laws about geoengineering. But he had read that the U.S. government ran war games to stop rogue geoengineers, and Wong's warnings had him worried.

As for the cooperation of the Mexican government, Iseman didn't think he needed it. By his own admission, even if his launches worked, the amount of cooling would be unnoticeable, a mathematical zero compared to the warming created by oil companies, none of whom sought his permission—or anyone's—to ruin the atmosphere. "I want to be launching fucking balloons and building shit!" he bellowed. "Not navigating fucking politics!" They called off the launch.

Iseman's project raises important questions about fighting climate change. Who will benefit from the new technologies to slow warming? Will the use of these technologies continue the historical pattern of entrenching inequalities and benefiting the elite? Or will humanity find a way to protect the most vulnerable and exposed members of global society? At the heart of these questions lies another one about governance: Who will steer mitigation efforts? Will it be entrepreneurs like Iseman? Will it be the governments of rich nations like the United States? Or will all the countries across the globe come together to slow climate change for the benefit of all? It is the hope of some that a democratic balance can be struck. Shade is a resource like any other. Its benefits will accrue to those who control it.

Shade on the earth's surface can help us adapt to the heat of climate change. It can cool human bodies, city streets, and entire regions. But because of rising concentrations of greenhouse gases in the atmosphere, less of the solar radiation that shade blocks leaves our planetary system. The infrared heat is absorbed, reemitted, and bounced back and forth between the atmosphere and the ground. What this terrestrial shade cannot do is change the temperature of the earth itself. Trees offer a notable exception to this rule, though not in the shade they cast. The moisture they evaporate can become clouds that reflect sunlight, and the carbon dioxide they sequester in their trunks allows more terrestrial heat to escape to space, thus cooling the planet. To make shade that does not just help us adapt to the rising heat of climate change but helps us slow it down, we need shade on a higher plane—in the atmosphere, or even in space, where it could block and reflect a small fraction of the sunlight that is the ultimate source of all planetary warming. This is what some geoengineers would like to do.

In 1989, a physicist named James Early drew up a plan for a giant, 1,240-mile-long sunshade that would live at Lagrange point L1, a gravitational position between the sun and Earth, about a million miles from home. To offset the warming forced by carbon dioxide and to cool the planet to its preindustrial climate, such a solar shield would only have to block less than 2 percent of the light that reached Earth's surface. Early's sunshade never came to be, not least because an object so heavy could not break through the earth's atmosphere and would have to be launched from the moon. In 2006, Roger Angel, an astronomer at the University of Arizona, came up with a way around this difficulty. What about sixteen trillion semi-opaque film screens launched deep into space? The screens would be operated by solar-powered ion thrusters, the same technology NASA uses to control satellites, and corralled into a sixty-thousand-mile-wide cylindrical cloud, also about a million miles away. Angel estimated twenty million metric tons of these robotic shades would be needed. Even deconstructed, the screen was too heavy for rocket fuel. Instead, a giant electromagnetic gun embedded in a mountain would fire the shades into space continuously for a decade. This plan, too, was shelved. Then in 2023, Morgan Goodwin, the director of the Sierra Club's Los Angeles chapter, resurrected the idea through a group called the Planetary Sunshade Foundation. Goodwin had spent his career fighting for a more livable planet and had come to see that earthbound efforts would not be enough. A giant solar sail, measuring nearly four hundred thousand square miles and weighing as much as one hundred thousand International Space Stations, could be his next step. "Skeptics might tell you there is no way of reversing global warming, but then many have said we would never walk on the moon," said Daniela Rus, a designer of yet another space shield, a raft of inflatable silicon spheres called Space Bubbles, in a TED Talk.

While these schemes are fantastical, the basic premise of blocking sunlight to shade the planet holds promise. The materials will not be Space Bubbles but chemicals like sulfur dioxide, calcium carbonate, aluminum, or diamond dust, which are easily injected into the atmosphere and known to reflect the sun's rays. As policies to limit emissions continue to fall short, and the dire forecasts of the U.N.'s Intergovernmental Panel on Climate Change (IPCC) arrive sooner than expected, a growing number of climate scientists and politicians are beginning to take a close look at planetary shade.

Atmospheric shade falls under a subset of geoengineering that climate scientists call solar radiation modification (SRM), the umbrella term for the new techniques and technologies being developed to cool Earth by reflecting the sun. Space shades are one example of SRM. Some geoengineers believe that brightening marine clouds (MCB) may effectively bounce sunlight into space. Raising the albedo, or reflectivity, of the many roofs and roads on the earth's surface is a project already under way. The controversial technique that Iseman pursues is called stratospheric aerosol injection (SAI), which proposes to send tiny particles into the upper atmosphere to dim the light, and thus the solar heat, of the whole planet.

The implications of SRM are enormous. By reducing the amount of sunlight absorbed by Earth, we may be able to avoid the worst effects of global warming. Animal species on the brink of extinction might return. Bleached coral reefs may be reborn. On land, the suffering of millions of people who would burn up in cities or starve in droughts would be avoided. SRM might slow the melting of the polar ice caps and turn back the rising seas. Some scientists even hope to restore a preindustrial climate.

This is the best-case scenario. But other scientists warn of doom. A failed geoengineering experiment could transform our environment in unexpected and undesirable ways. Under a modified atmosphere, monsoons that countries rely upon for farming could disappear. Atmospheric disturbances such as heat domes and hurricanes could become more common. The truth is, we have no idea what would happen, which is why many scientists and policymakers are adamant that we should never try it. They believe SRM offers an awesome power that humanity is incapable of responsibly managing, so it would be better not to consider it at all. "There are many people who are out there, especially if it's not them who will die, that will say that it's so risky to mess around with nature," said Janos Pasztor, a former U.N. climate diplomat who played critical roles in the formation of the United Nations Framework Convention on Climate Change (UNFCCC) in 1992 and the negotiations that led to the ratification of the Paris Agreement of 2015. Today, he advocates for responsible governance of SRM technologies.

A commonly held objection to SRM is that it treats the symptoms of climate change without addressing the underlying cause of carbon emissions. This is often accompanied by a visceral disgust at more pollution being the solution to pollution. Instead of continuing our difficult work of getting off fossil fuels, these objectors worry, geoengineering would allow us to continue business as usual. Although SRM would combat the heating effects of carbon dioxide, air quality would continue to worsen and so would ocean acidification, threatening marine life. The concern with geoengineering's "moral hazard" is not rank speculation. In 2008, Newt Gingrich, the former U.S. house speaker, pushed geoengineering as a cheap alternative to reducing emissions. These objectors argue that anything that could undermine the resolve to cut emissions should be avoided. But now that the window has practically closed on the opportunity to meet the Paris Agreement target of halting the global temperature rise at 1.5 degrees Celsius, the mood has shifted. "The question is not whether it's more likely than not to overshoot, but how long will the overshoot last, and how high will the temperature be," Pasztor told me. Thanks in part to the work of people like him, humanity has avoided the so-called business-as-usual scenario, which could have warmed the planet by 4.3 degrees Celsius (almost 8 degrees Fahrenheit) above the historical average by 2100. Even so, Pasztor believes the planet is still likely to warm by 2.6 to 3.1 degrees Celsius (around 5 degrees Fahrenheit), an outcome that would still be catastrophic.

In unison, geoengineering researchers say the number one priority is ending greenhouse gas emissions. For this, there is no substitute. But David Keith, a University of Chicago geophysical scientist who strongly advocates for SRM, believes the idea that climate change can be solved with emissions cuts alone is a "dangerous fantasy." Keith asks us to suppose emissions will be eliminated by 2050. Average temperatures will stop rising, but he says cooling will take thousands of years as carbon dioxide slowly leaves the atmosphere. Heat waves will remain fatal, polar ice will continue to melt, and the seas will rise. By then, direct air capture systems and carbon dioxide scrubbers could be pulling heat-trapping gases out of the air, but success is far from guaranteed. Unless these machines are miraculously efficient, some cities could still become uninhabitable and entire societies could drown. In such a scenario, a desperate public will demand action. If we want to avoid this outcome, SRM will be necessary.

Global Shade Experiments

Though many of us see them as little more than harbingers of rain, clouds are hugely influential in global weather. For one, they help control the temperature. Clouds cover two-thirds of the planet's surface, and depending on how high and thick they are, and the amount of water and ice they contain, they either shade it by blocking sunlight or warm it by trapping terrestrial heat. The clouds that provide shade are marine stratus and stratocumulus. These billowing duvets painted by Georgia O'Keeffe and glimpsed through windows on flights bounce 30 to 70 percent of the sunlight that strikes them back into space. The heat trappers are the wispy, silky cirrus clouds we see from the ground. Global warming could change how clouds form and, in the worst-case scenario, rob our planet of its natural sunscreen forever. On the other hand, making clouds more reflective could keep the planet cool. Since the 1990s, scientists have suggested that cloud albedo could be increased by spraying seawater into the air from boats. At scale, one scientist suggested that a mere 3 percent increase in reflectivity could offset the warming effects of CO_2. A version of the "albedo yachts," as one scientist called them, is now in action in Australia.

Brightening the clouds in sensitive areas such as the Arctic could have globally beneficial effects. University of Cambridge cloud brighteners imagine a fleet of hundreds of vessels near the North Pole, firing off plumes of seawater to shade the thawing ice. But MCB may be more viable as a regional solution. On the other side of the planet, Australians have since 2020 ferried research vessels to the Great Barrier Reef to fog the air above the beloved corals that are bleaching during unforgiving marine heat waves. On such a mission, Southern Cross University scientists pointed cannon-shaped turbines to the sky and shot trillions of droplets of ocean water into the air. The salty mist caught a warm draft and floated into the atmosphere, where it thickened the clouds that shaded the coral. Daniel Harrison, the scientist leading the Reef Restoration and Adaptation Program's Cooling and Shading R&D Subprogram, found the fog blocked an additional 30 percent of incoming sunlight, which delayed the bleaching of some coral for three weeks. The initial results are encouraging, but to have a meaningful effect on even a small patch of coral, Harrison would need ten times as many boats. And to shade the entire Reef, the scientists would need eight hundred boats spraying continuously for months ahead of expected heat waves.

These trials are only one element of Australia's much larger billion-dollar campaign to protect the Reef from climate change, which includes attempts to breed heat-tolerant coral. In one significant aspect, the Australians have already succeeded. They have escaped the public backlash that has torpedoed other geoengineering efforts, like Iseman's forays in Mexico, and more conventional experiments, like those of the climate NGO SilverLining. In 2024, SilverLining had attempted to fund a project to study cloud brightening from a decommissioned aircraft carrier parked in Alameda, California. Because there are no laws around geoengineering, the science team didn't have to inform the local authorities about the true nature of the work. When Alameda's city council learned that the "atmospheric science educational exhibit" was in fact a geoengineering field experiment, they moved quickly to shut it down.

Possibly, Harrison and the Australian scientists encountered less resistance to cooling and shading because they pursued a popular goal with a clearly defined outcome: Save the Reef. The Reef is a closely managed conservation site that is protected by legislation and a special federal agency. Research was only permitted after community consultation and extensive public input. And before they entered the field, the researchers sought and gained common ground with the Aboriginal and Torres Strait Islander groups that Australia recognizes as the Reef's traditional owners. Harrison's technical achievement is not as impressive as his social achievement. Where the scientists in Alameda found themselves at odds with democratically elected leaders, Harrison was able to fit his work into a broader public project.

Surface-bound efforts like Harrison's can reach the clouds in the troposphere, the lowest layer of the atmosphere, and give them a reflective boost. Other scientists are thinking about SRM on a higher level. Their sights are aimed at the stratosphere, the atmosphere's second layer, where commercial flights reach cruising altitude.

The first to propose stratospheric aerosol injection (SAI) was a Russian climatologist named Mikhail Budyko. In the 1960s, Budyko developed a global energy balance model that estimated how much of the solar radiation that struck Earth was reflected, absorbed terrestrially, or emitted back to space. For most of human civilization, this balance was just right. The most abundant greenhouse gas in the atmosphere, water vapor, trapped enough outbound radiation to warm the surface of the planet to a mean temperature of 59 degrees. But the scales began to tip in the eighteenth century, as carbon dioxide, methane, and nitrous oxide began collecting in the atmosphere and storing more heat on Earth. Budyko was among the first to accurately predict the terrible consequences of this radiative forcing: rising global temperatures and melting polar ice. In a 1974 study, he suggested that the simplest way to cool the planet would be to prevent some of that radiation from entering in the first place. A stratospheric shroud of sulfuric aerosols might do the trick.

It wasn't until 2006 that someone took Budyko's blanket seriously. In a now-famous essay, Nobel Prize–winning chemist Paul Crutzen zeroed in on an unintended consequence of progress. The U.S. Clean Air Act and its various amendments had improved air quality and saved countless lives by scrubbing the atmosphere of sulfur dioxide (SO_2), the filthy chemical compound emitted from coal-burning power plants and smelters. But in doing so, the United States had cleared the skies, allowing more sunlight in to warm the planet. We rightly understand that sulfur dioxide makes our lives worse, causing asthma attacks and acid rain. But when the gas enters the atmosphere, it mixes with water vapor to form aerosols that reflect sunlight and shade the planet, cooling the global surface by about a degree. In fact, if it hadn't been for pollution, the warming we experience today would have begun decades earlier. In recent years, scientists like James Hansen, who sounded the alarm on the greenhouse effect to the U.S. Senate back in 1988, have observed an escalating rate of warming. They believe that cleaner air and clearer skies may be the cause.

Since the U.N.'s Kyoto Protocol of 1997, reducing emissions has been accepted as the best way to stop global warming. Crutzen believed this was a "pious wish." He offered Budyko's blanket as the realistic solution. The increasing amount of solar radiation that reaches Earth, he believed, could be diminished by the regular injection of sulfur particles in the stratosphere. In other words, he wanted to move industrial pollution higher in the atmosphere. Crutzen thought the amount of sulfur needed to create a cooling effect would be so small that it would not contribute to acid rain or harm human health. There would be some negative impacts, of course. The risk of skin cancer might increase because sulfuric acid corrodes ozone. And the sky would look a little different. The aerosols that bounce solar radiation back into space also scatter light waves in the atmosphere. Right now, the daytime sky is blue, but with more aerosols, it could be white. The glowing halo we see around the sun could appear larger. And at sunset, the colors could be stronger and more vivid, boasting brilliant red afterglows. But those side effects would be insignificant compared to the benefits of slowing global warming.

Crutzen wasn't sure how to get the particles up there—maybe with airborne artillery, maybe with battleship guns, or maybe with weather balloons that exploded in the sky, the delivery system preferred by Iseman. He was certain, though, that stratospheric sulfur would work, because it had before. In 1991, Mount Pinatubo, in the Philippines, erupted and blew eleven million tons of sulfur dioxide into the stratosphere. For six months, the skies above the volcano remained dark. By month fifteen, the shroud had circulated around the globe, which cooled the earth's surface by about 1 degree, temporarily undoing fifty years' worth of man-made heating. Not long after, the aerosols drifted to the poles and descended in rain droplets. The skies turned blue once again, and global temperatures continued their inexorable climb.

Other shorter-lived bursts of cooling have occurred after natural disasters. Scientists observed an effect after wildfires in Australia in 2019 and 2020, when an updraft punched a million tons of smoke into the stratosphere, reducing global temperatures by one-tenth of a degree. The smoke cooled the planet more effectively than a full year of pandemic lockdowns, in part because the decrease in carbon emissions was offset by the decline in planet-shading pollutants. Locally, the cooling effects are even stronger. In 2023, as smoke from Canadian wildfires shrouded parts of Pennsylvania and New York in ghastly orange skies, one pundit declared the 9-degree temperature swing the silver lining of Smokemageddon.

Short of setting the planet on fire, the best method of particle injection would be with planes. A fleet of aircraft cruising at sixty-five thousand feet and spraying twenty-five-ton payloads of sulfur dioxide would get the job done more efficiently than millions of exploding weather balloons. A Columbia University climate economist named Gernot Wagner and a former airline executive named Wake Smith have suggested "stratospheric aerosol injection lofter" planes that could be developed for around $2 billion and flown continuously for a decade. All told, it might cost $36 billion to cool the planet for fifteen years, which isn't even that expensive. It's less than what the U.S. Department of Defense spends on building new fighter jets each year and much cheaper than the $100 trillion it may cost to end our dependence on fossil fuels.

SAI is cheap enough that motivated geoengineers can easily begin cooling unilaterally without access to a military budget. Wagner imagined that decentralized solar geoengineering could be initiated by hobbyists who buy balloons, sulfur dioxide, and helium canisters online. Progress would be painfully slow. An amateur geoengineer would have to do this millions of times to lower the global temperature by one-tenth of a degree. But it was only a matter of time, Wagner wrote, until someone decided to try. If that someone was not Luke Iseman, but a climate-conscious billionaire, the stakes would rise. Pasztor warns that Elon Musk or someone of his ilk could easily bring geoengineering to scale and meaningfully change the global climate without scientific oversight or democratic control.

SAI has the potential to change Earth's atmosphere in irreversible ways. For this reason, it is critically important to take unintended side effects into account. The computer models that show how stratospheric aerosols can block sunlight and lower temperatures also show that they could slow the jet stream and create more heat domes. They could stimulate more hurricanes or end monsoons. No one can say for sure what will happen. The worst outcome would be a geoengineering scheme that amplifies the burdens of climate change. English computer scientists used models to simulate the effects of aerosols sprayed over the Northern Hemisphere. This would have a strong cooling benefit for the United States and decrease the severity of Atlantic hurricanes. But at the same time it would have a devastating impact on Africa by intensifying drought in the Sahel. It's not just conjecture. Smoke from the Pinatubo eruption weakened the seasonal monsoon that Indian farmers depend on.

In a highly unequal world, the burdens and benefits of stratospheric shade are likely to be unevenly distributed. Who's to say that another disturbance, one that is the result of human intervention rather than volcanic activity, wouldn't threaten millions in the Global South with starvation? Pablo Suarez, the late Red Cross humanitarian, worried that "we let them eat the risk that we create." Pessimists believe it would be impossible to arrive at a fair international agreement about geoengineering. More than five hundred academics have signed and two thousand NGOs have backed a "non-use agreement" and declared that since SRM cannot be governed impartially, national governments and the U.N. should issue a complete moratorium.

If a mass injection of stratospheric aerosols does more harm than good, or a worldwide deployment creates climate winners and losers, it would be better to know that now before someone makes a rash decision. Geoengineering boosters readily admit that SRM could be dangerous. But so is climate change. They seek to hedge against the risk of unintended consequences by conducting small-scale but real-world field tests to build data.

Until a few years ago, SRM research occurred only on computers.

Scientists who hoped to conduct outdoor experiments could find no government support for their activities. Privately funded university trials were canceled under pressure from activists, who claimed that experiments were definitionally a form of deployment, because they were, however infinitesimally, modifying the climate. They were adamant that the only climate modification should come in the form of reduced emissions.

In 2021, David Keith, then at Harvard, and a colleague named Frank Keutsch hoped to travel to Kiruna, Sweden, a small town above the Arctic Circle. As part of the Stratospheric Controlled Perturbation Experiment (SCoPEx), they planned to loft an elaborate aerosol sensor into the lower stratosphere and release about two thousand grams of calcium carbonate. The sensor would fly back and forth through the chemical plume, recording the reactions of those aerosols with other atmospheric particles. The calcium carbonate would reflect an insignificant amount of sunlight, less than sixty seconds' worth of condensation trails from a Boeing 747. But three months before launch, the scientists began facing pushback from the Sámi, the region's Indigenous inhabitants. They were joined by Sweden's most influential climate activist, Greta Thunberg.

The Sámi have traditionally been reindeer herders. Their livelihoods depend on what happens in the skies above Sweden. As the Arctic warms, more snow is falling as rain, which in the winter forms a thick ice crust that prevents the reindeer from grazing. As a result, reindeer are starving to death. Even though SRM could potentially save their way of life, the Sámi asked the Swedish government to halt the Harvard experiment. "There is a need for technology development," explained Åsa Larsson Blind. "This is not something that we question at all." Larsson Blind is a politician on the Saami Council who became a leader for the opposition. "What we do oppose is the direction toward technologies that do not actually target the root causes of climate change." Rather than "a quick fix, and the overbelief of technology being the main solution," Larsson Blind urges a transition "toward a more sustainable way of living."

On the advice of an independent advisory committee, Keith decided to postpone the launch until Harvard could come to an agreement with the locals. But that never happened, and the project shut down for good in 2024. Keith was frustrated. "There are lots of examples of these moral hazards or risk compensation in public policy and they don't mean we shouldn't do things. There were people who argued strongly that we should not put airbags in cars because they would encourage more dangerous driving," he later argued. In another interview, he said of SCoPEx, "If we do this again, we won't be open in the same way."

The conflict with the Sámi points to a bigger question: Who gets to make decisions about the future of the planet? Whose hand should be on the global thermostat? Given the dearth of public support for SRM research, projects have had to rely on private funding. David Keith's work, for example, was backed by Bill Gates. George Soros has also taken an interest in geoengineering experiments. "Private philanthropy money seemed to us the only option," wrote Gernot Wagner, who co-founded Harvard's solar geoengineering program with Keith. This means that decisions affecting the course of human history could be made not by publicly accountable governments but by private individuals.

So far, the most prominent of these individuals has not been a scientist concerned with building knowledge like Keith, but Luke Iseman, an entrepreneur governed by the profit motive. In addition to amassing a customer base and a coterie of online enthusiasts, Iseman has told the world about geoengineering in the pages of *The New York Times,* on NPR broadcasts, and on CBS daytime news. The unflappable Iseman spent his twenties and thirties hawking sustainable products he hacked together, like pedicabs, Wi-Fi garden sensors, and off-grid shipping container houses. He didn't know much about SRM until he read *Termination Shock,* a science fiction book about a Texas oil entrepreneur who builds a giant cannon to shoot aerosols into the sky. In a flash of inspiration, Iseman landed on his next venture. It seemed like this tech was simple and cheap. Maybe there was a customer base for it.

In April 2022, Iseman tested two homemade sulfur dioxide–filled weather balloons, launching them into what he claimed was the stratosphere. He planned to sell every gram of reflective material as a $10 cooling credit, which, according to his calculations, offset one ton of carbon emissions. Compared to other forms of carbon offsets, such as forest credits and carbon trading, his cooling credits were a bargain. So what if there was no way to prove the credits actually cooled the earth? Iseman convinced two venture capital firms to fund him at a modest $750,000 to found the start-up Make Sunsets. Three years later, Iseman's company and weather balloons are still afloat. (I was an early customer.) He claims to have sold enough credits to offset two years' worth of pollution by the island nation of Kiribati.

There are reasons to be troubled by Iseman's success, however modest it may actually be. The first reason is that unverified cooling credits could do more harm than good. Scientists who value precision and accuracy believe that Iseman's measurements are too casual. Since his interest is in selling his sulfur launches as cooling credits, he is not incentivized to look too closely into whether they work. "There's uncertainty about all of this, but there's not uncertainty about the fact that putting sulfur dioxide into the stratosphere creates cooling," Iseman told me, the first time we talked. "That is very, very clear." He beseeched me to look at the big picture. According to Keith, the University of Chicago scientist, the big picture is that commercial enterprises like Iseman's have financial motives to oversell the benefits, downplay the risks, and continue selling their services, regardless of the impact. Iseman's clients are free to believe they are cooling the planet, whether they are or not, giving them cover to continue polluting.

Scientists do not believe the difficult and high-stakes work of geoengineering should be left to amateurs like Iseman. Not long after the Mexican launches, Jesse Reynolds, then an official on the Climate Overshoot Commission, a project of a Paris-based NGO that built support for SRM research, interviewed Iseman on a podcast. "I'm going to speak bluntly here," said Reynolds. "Less than a year ago, you learned about SRM, and you spent maybe, I don't know, fifty bucks on Amazon and bought a couple of balloons and bought some elemental sulfur that you burned into what you concluded was SO_2, got it into the balloons, and let them go. Didn't measure where they went. Didn't measure how high they went." The accusations piled up. Iseman didn't file a patent. He exaggerated the cooling effect. He drew buyers away from legitimate carbon markets. Given all that, Reynolds asked Iseman, "Why should the SRM community, broadly defined, take you seriously?"

Silicon Valley's move-fast-and-break-things ethos embodied by Iseman has made people wealthy. But when the thing in question is the planet, breaking it has far-reaching consequences. Iseman's solutions elide the need for public input into a public problem. Perhaps unsurprisingly, tech giants are among the biggest investors in geoengineering, having plowed millions into efforts to test and model its potential impacts. They want to bypass the difficult work of decarbonization with the one weird trick that will solve climate change. They believe they can "unfuck the planet," as geoengineering investor Chris Sacca put it, without engaging in the messier work of social change and public accountability. "Legitimate, well-governed research in solar geoengineering is important. That is not what Make Sunsets is," wrote Shuchi Talati, a former Biden administration energy official, on the social media platform X. "At best it's a stunt, but at its core it is anti-democratic: they are working to take away the agency of people to participate in decisions that will deeply impact them."

"Can any of you explain to me one innovation that has happened after asking permission?" Iseman asked me. Sitting cross-legged on a yoga mat, he compared himself to Edward Jenner, the eighteenth-century English doctor who experimented on living volunteers during the successful invention of the smallpox vaccine. At that time, smallpox had been a scourge on humanity for thousands of years, and would go on to cause hundreds of millions of deaths. Iseman mentioned a new study that suggested the lives lost to climate change could one day eclipse smallpox mortality. "It's morally wrong, in my opinion, to have this tool that no one debates whether it would save lives," and not use it, Iseman said. But he was incorrect. There is a debate about the ability of stratospheric aerosols to save lives. His decision to launch without seriously considering the unwanted side effects in countries where SRM could be ruinous calls attention to an important dimension of the geoengineering debate: inequality.

The Dream of Just Deliberation

As the United States, the United Kingdom, and others accelerate SRM research and policymakers contemplate paths to deployment, the countries with the most at stake are just getting started. The Degrees Initiative, an English NGO funded by Western charities and donors, now sponsors scientists in Africa, Asia, and Latin America, regions where billions of people are on the front lines of climate change, to run SRM models and probe the impact on their countries. The goal is to build local knowledge and awareness of SRM, even as the global superpowers remain miles ahead, or more accurately miles above, in their progress. Indeed, the prospect of one country rapidly developing an SRM program to the point of deployment while others are still catching up to the ramifications troubles many in the geoengineering space. They believe that a globally impactful technology should be globally governed.

Shuchi Talati hopes that governance could be led by countries whose fate hangs in the balance. Since 2023, she has pursued this goal as the founder of the Alliance for Just Deliberation on Solar Geoengineering (DSG), a nonprofit that focuses not on the technology of SRM but its management. In the United States, there are climate entrepreneurs, D.C. policy shops, and federal officials who are contemplating if SRM could be deployed and by whom. Such discussions are rarer in countries where DSG has worked, such as South Africa, India, and Pakistan. The science conversations are happening, but the governance conversations are limited.

Because of groups like the Degrees Initiative, atmospheric and meteorological scientists are now investigating the unequal climate implications of a more reflective global atmosphere. But governments and civil society groups are only dimly aware of SRM, which makes it difficult for the public to grasp its costs and benefits. In fact, said her colleague Hassaan Sipra, many people are still in the dark about the implications of climate change. The notion that there are options to fight it besides simply reducing emissions is startling. DSG organizes workshops at think tanks and universities to educate governments, environmentalists, human rights advocates, and other climate stakeholders about SRM. The aim is to give them the tools to make informed decisions about the technology, including whether they even want to use it. In the case of Pakistan, DSG partnered with an Islamabad-based sustainability expert to publish a brief with a set of recommendations, such as folding SRM into the country's official climate policy, and investing in regional climate models to better understand, for example, how a change to the summer monsoon could affect the water supplies the country shares with India. Armed with such knowledge, they urge Pakistani negotiators to ramp up their presence in talks about SRM on the international stage.

Others allege that work like Talati's is the tip of a dangerous spear. They believe that such workshops manufacture consent for a Global North technology that the Global South has not asked for and in some cases has actively rejected. At a 2024 panel in New York about geoengineering, Carol Bardi, a Brazilian environmentalist who organized the non-use agreement to ban SRM, pointed out that two thousand global organizations have already registered their opposition. Calls from the Global North for more engagement ignore the fact that many in the Global South have already weighed in. "When to stop? What is a no?" she asked.

Talati and her supporters see the situation differently. The technology is already here, and no online petition can stop its momentum. If important decisions about SRM were made today, the conversations would be steered by American, European, and Chinese interests, and even entrepreneurs like Bill Gates and Luke Iseman. Could they really be trusted to look out for the well-being of billions of people in South America, Africa, and Asia? Talati thinks not. As part of her U.S. Department of Energy portfolio, Talati oversaw the development of carbon dioxide removal, a powerful tool to pull greenhouse gases out of the air. The technology looks promising, but it will still be many years before it can change the global climate for the better and offer humanity any relief. By comparison, SRM is fast, cheap, and can be launched unilaterally, right now. In her estimation, the government was not keeping pace with the rapidly evolving field. A major concern was that SRM could become the province of the private sector. "I just felt like, it's now or never," Talati told me, of her decision to leave her post in the Biden administration to focus on just deliberation. "There's a very narrow window of opportunity to do this work for SRM in a way that it can actually change the dynamics of this field."

Climate decisions made by individual countries have global ramifications. If any single nation decides to fire reflective particles into the stratosphere to shield itself from the heat of the sun, the aerosols will swirl around the globe and dim everyone's light. Truly just deliberation would require the cooperation of every country on Earth, and for that reason, it is difficult to imagine how it could proceed. There is no real precedent for global collective action on this scale. The U.N. would seem to be the natural venue for such an attempt. In 1987, nearly two hundred countries unanimously ratified the Montreal Protocol to ban the chemicals in refrigerators, air conditioners, and spray cans that deplete atmospheric ozone. The ozone layer is on track to fully recover by 2066, less than a century after this treaty was signed. In 1992, one hundred and fifty-four countries and the European Union signed the UNFCCC treaty to stabilize the concentrations of greenhouse gases in the atmosphere, and the subsequent Kyoto Protocol and Paris Agreement set goals for lowering emissions. Although the growth of emissions has slowed, countries are still polluting and the planet is still warming.

I asked Pasztor to imagine how an ambitious agreement on global cooling could take shape. He speculated that the U.N. General Assembly could initiate the process by deciding to coordinate international research on stratospheric aerosols that fully assesses their risks and benefits. In this hypothetical scenario, it could take at least ten years to gather enough data to make an informed decision on whether to proceed. At that point, the member states could weigh the costs of ecological destruction and human suffering against the projected effects of SRM. If they decide that the risks of aerosol injection are greater than the risks of inaction, then their conversation will end. But if they decide that millions of deaths and expensive adaptation efforts—air-conditioning, seawalls, and mass relocation—are unacceptable, then the countries will begin the difficult work of establishing an international authority to oversee the continuous injection of aerosols into the stratosphere. In two or three generations, when the concentration of CO_2 in the atmosphere finally peaks, the operation could wind down. In Pasztor's scenario, stratospheric aerosols, if ever used, would be a temporary corrective, and on the scale of the thousand years it will actually take for carbon dioxide to leave the atmosphere, a brief one.

Pasztor sees the International Atomic Energy Agency, a U.N. organization that oversees nuclear nonproliferation and its peaceful use for power generation, as a precedent. Even so, a geoengineering treaty could require even greater international coordination to actively build something new. The world has never seen this form of cooperation. According to Pasztor, it could be the most challenging global task ever.

Although the U.N. may offer the best chance for global oversight, Pasztor has pursued different opportunities to bring SRM to the attention of world governments. From 2017 to 2023, as the executive director of an independent climate project, once known as the Carnegie Climate Geoengineering Governance Initiative, he met with hundreds of government officials and NGOs, urging them to come to grips with the realities of climate-altering tech. Even then, it was nearly certain that the goals of the Paris Agreement, which he had been instrumental in securing, would not be met. He encouraged them to take SRM seriously by investing in research and creating policies for local governance. But progress has been slow. The predictable result of government inaction has been a legal vacuum where entrepreneurs like Iseman are free to make a buck on the back of the planet. Pasztor considers Make Sunsets, with a market cap around a million dollars, a mere provocation, and knows that Iseman's launches have no physical impact on the atmosphere. But Pasztor also believes that Iseman heralds more serious problems. If SRM remains unregulated, then billionaires answerable only to investors and their own egos could easily succeed in radically transforming the global climate.

"Wait until Mr. Musk does this for ten billion," Pasztor warned. What worried Pasztor was the leverage the world's richest man would have over governments if he unilaterally injected aerosols and the world was forced to rely on him. "Let's say a private individual who is accountable to nobody starts doing this for a few years, and then says, 'Okay, I will stop, unless you give me a tax break,'" he continued. "This is the kind of stuff that could happen." That person would have the power to trigger "termination shock," the sudden rise in temperature that could occur if the planet abruptly lost its global sunscreen. Pasztor believes a global governance framework could protect against such a catastrophe.

In February 2024, for the second time in five years, the 190-plus member states of the United Nations Environment Assembly (UNEA) considered a Swiss resolution to lay the groundwork for global management of SRM. It called for the United Nations Environment Programme to assess the state of the science and take stock of the activities now popping up around the world. Even this small step proved too fractious. Once again, the members failed to reach consensus. That the countries could not come together to do anything at all about solar geoengineering deeply disturbed Pasztor. The world was on the precipice of a long and painful temperature-overshoot scenario. Entrepreneurs like Iseman were undertaking the risky work of modifying the atmosphere without oversight or transparency. The paralysis was indicative of larger problems—growing political divisiveness, international mistrust, a lack of faith in institutions, and the inability of self-interested countries to pursue global cooperation.

Three months later, Pasztor announced that he had joined Stardust Solutions, an American-Israeli solar geoengineering start-up, as an independent consultant. The company said it had $15 million in funding to start conducting tests to eventually launch aerosols into the stratosphere. Pasztor, once the U.N.'s great climate champion, was slammed for working with private geoengineers, even though he had his earnings donated to the United Nations Relief and Works Agency for Palestine Refugees. But he felt he had no choice. "Stardust exists. Investors are funding it," he wrote. "Doing nothing is not really an option." Hopefully, someone will one day shade the planet responsibly.

On February 12, 2023, I was freezing in a windy park in Reno, Nevada, trailing Luke Iseman, Andrew Song, a three-person film crew, and a reporter and photojournalist from *Time* magazine. After the imbroglio in Mexico, Iseman had relocated the Make Sunsets operation back to the United States, where he hoped to be free of government interference. We were there to attend what Iseman said would be the first-ever injection of stratospheric aerosols from American soil.

Iseman assured us he was cleared to launch. He had filed notice with the FAA, which controls U.S. airspace, though he neglected to mention that his balloons were filled with sulfur dioxide. Iseman had been contacted by the FBI, but it wasn't about dangerous chemicals. The entire country was on edge about mysterious balloons in the sky. The military had shot one down, and the feds wanted to know if any other airborne objects they were anxiously tracking belonged to him. Iseman was amused but not concerned. U.S. national security, like the Mexican environment, was overshadowed by the imperative to move fast. "When people scream, it's because it's important, and you are challenging some piece of their worldview," one of their investors assured them, over email. "Just keep building. Gonna be fun."

Iseman and Song's operation had not changed much since Mexico. They huddled around a barbecue grill, burning sulfur powder and storing the smoke in a garbage bag. Families watched Song hacking on poisonous clouds. "So fucking amateur," chuckled Iseman. Song transferred the smoke from the garbage bag to a weather balloon. Rather than weigh the SO_2, Iseman and Song guesstimated one balloon's contents to be a little less than ten grams' worth of cooling. Or maybe five.

As the sun set behind the Sierra Nevada, I helped Iseman and Song drag the smoke-filled balloon, a wooden board festooned with tracking equipment, and a 150-pound helium tank to a grassy clearing. Sitting astride the tank, Iseman instructed Song to carefully hold the balloon as he inflated it with helium, a few hundred grams at a time. The limp balloon swelled to a nine-foot orb. As Song held the balloon's neck, Iseman removed the inflator tube and cinched it shut with Gorilla Tape and a rubber band. Then Iseman took the balloon from Song. He would be the one to launch.

The cameras were rolling. A drone whirred overhead. Iseman took a few steps into the wind, holding the balloon and its trail of tracking equipment. At 5:17 P.M., looking to the sky, he let go. As the balloon bobbed into the heavens, the filmmakers prodded Iseman for a Neil Armstrong–esque declaration. "Are you crossing a threshold?" one asked. Iseman did not answer. He was fixated on the balloon as it faded into the atmosphere. Whatever happened next was out of his hands.

Epilogue

We all act out of self-interest. This is natural, and according to some, desirable. Yet even the most seemingly benign of our individual decisions can have unwanted consequences. When we build an asphalt driveway for our new SUV, we reduce Earth's albedo and absorb more heat. If we cut down a tree to build that new driveway, our neighborhood becomes a little brighter, the air slightly drier, and the ambient temperature imperceptibly higher. And when billions of us do these things, the small consequences amount to a problem far beyond the ability of any one person to solve. Climate change must be addressed at the scale of culture, society, and the environment—ultimately, as Janos Pasztor suggested, at the scale of humanity itself.

When Los Angeles residents chill out in air-conditioning, they feel comfortable and stay healthy. This personal comfort comes at the cost of urban heat and carbon emissions. But because of the way L.A. is designed, its buildings engineered, and its streets optimized for cars, Angelenos often have no other way to cool off. We cannot simply ask them to suffer in the heat. We need public planning to build the homes, sidewalks, and parks that keep them cool and healthy without AC. We need a lot more shade.

Unfortunately, we all have to go to work, and most Americans can only get there in a car. Although vehicle emissions warm the planet, we can't simply ask everyone to stay home come Monday morning. We need to build the transportation infrastructure that makes it easier to walk, bike, and board the bus. Giving up the little ACs in our cars will require shade.

Even our solutions to climate change are predicated upon individual action. We can all plant trees. We can all buy carbon offsets when we fly internationally. And now, thanks to Luke Iseman, we can buy cooling credits. Floating balloons into the heavens one by one assuages guilty consciences and makes money for Silicon Valley entrepreneurs. But for geoengineering to benefit Burkinabe cattle herders and homeowners in Phoenix, it will require global cooperation.

This book has introduced you to people doing their best to adapt to a warmer world. Tony Cornejo built a little shelter for his neighbors waiting outside. In the fields of California, advocates work on behalf of the poorly paid farmworkers who have no choice but to power through dangerous heat. Debbie Stephens-Browder planted trees to cool the streets that were her only home.

Others hope to design a fairer world. In Portland, Oregon, the professor Vivek Shandas called our attention to the unjust ways cities are built. The neighborhoods shaped by redlining in the twentieth century are red-hot in the climate-changed twenty-first. The architect Jeff Stern designed a home safe for inhabitants and the planet alike. In L.A., the Kounkuey Design Initiative snuck shade into the space spared by anachronistic land use regulations that tie our hands with red tape when we most need to hold a parasol. And climate entrepreneurs like Luke Iseman sincerely believe we can innovate our way to a better world. They are understandably growing impatient with the slow pace of progress.

But even if Stephens-Browder filled every tree well, her neighborhood would still be hotter than one on the other side of town. No matter how efficient passive houses get, they will not make an appreciable difference in emissions until millions of them are built. And even if there was a Sombrita on every sidewalk, the rest of the street would still be unshaded. Just as a small, perforated screen is insufficient shade on the earth, so too are a million balloons in the stratosphere insufficient to furnish shade in the skies.

No amount of individual creativity will ever solve a planetary problem like climate change. Climate adaptation requires more than sophisticated technology. To keep the earth livable, we need collective planning, management, and action. This is where shade comes in. In most cases, it does not require innovation or cutting-edge technology. It is a concept so simple that it is grasped instinctively by insects, lizards, and fish, and it scales. It could be one of our most powerful defenses against rising heat.

Effectively governing the global climate may be the most challenging thing humanity ever does. But we cannot be daunted by this task. Because if we do not find a way to act responsibly in the collective interest, then we will all be left to continue to travel our individual paths. And without a mass movement for climate adaptation and a collective mechanism that supports and directs it, these paths will grow more isolated. And hotter.

The conversation about shade has just begun.

Acknowledgments

I could not have written this book without the patience and support of my partner, Cybelle Tondu. For years, she offered insights and reassurances and believed in me when my own confidence flagged. After our son, Eli, was born, she took care of our family when I could not. Most important, she made sure I finished the book. Thank you, Cybelle.

My agent, Elias Altman, and my editor, Molly Turpin, were the book's champions. Elias worked with me to shape a jumble of ideas about shade into a coherent proposal, found the book a home, and continues to be my sounding board. I owe him a debt of gratitude for his advice, and above all, for his advocacy. Molly's enthusiasm never failed to lift me up, and when I was overwhelmed by the material, she did not lose sight of the book's essential curiosity and wonder. Her close reads and skillful edits helped me focus and find throughlines within and across the chapters, and she offered essential guidance on structure and length.

Special thanks to Monica Brown for shepherding the book through production and answering my questions about the minutiae, and to the entire team at Random House that brought the book to life: Edwin Vazquez, Ada Yonenaka, Rebecca Berlant, Richard Elman, London King, Marni Folkman, Michael Hoak, Andy Ward, Alison Rich, Ben Greenberg, and Erica Gonzalez. Thanks also to Emily Lavelle for increasing the book's visibility.

No work of journalism is possible without sources. My sincere appreciation for the time shared by hundreds of people I interviewed for the book—scientists, urban foresters, city planners, environmentalists, farmworkers, advocates, academics, engineers, civil servants, business owners, architects, doctors, politicians, activists, entrepreneurs, physiologists, and concerned citizens—and their repeated explanations, for continuing to talk and email with me when I needed more, and trusting me with their stories.

Between 2021 and 2023, my research, reporting, and writing was financially supported by Emerson Collective and MIT Knight Science Journalism, and grants from the Robert B. Silvers Foundation and Vassar College. At Emerson, thanks to the late Amy Low and Patrick D'Arcy for their interest in my project, Kevin Dupzyk for the constructive edits on a few chapters, and the other members of my fellowship cohort for the conversations, especially Jake Barton and K-Sue Park.

I am deeply grateful to Yan Slobodkin and John Palattella for their structural edits. Siri Chilukuri, Francis Carr, Jr., and Savannah Huguely for meticulous fact-checking, careful double-checking, and probing questions, all of which made the book better and instilled me with confidence. Any errors are mine alone. Elizabeth Gumport, for research and reporting assistance. Meredith Sadler designed skillful diagrams, and Adam Baron-Bloch photographed La Sombrita. Special thanks to Jason Fulford for lending the striking image that adorns the cover. In Los Angeles, thanks to Eva Grenier, Weston Rowland, Bret Nicely, and Sylvie Turk for assisting in crafting, filing, and reviewing public-records requests, and Ivan Horacio for translations. In Seville, Felicia Coffey, Lorena Garcia, and Eric Uguet were my able interpreters. Thanks also to Shula Smith for translating Hebrew documents, and Della Chan for Chinese interpretation in New York.

In the 2010s, two people opened my eyes to shade: James Rojas, who alerted me to its importance in the built environment, public and domestic, and the Latino impulse for shade-making, and Sahra Sulaiman, who chronicled the neglect of L.A. bus stops through dramatic photography and incisive reporting for *Streetsblog LA*. James has been a reliable interlocutor of shade ideas and observations, and for this book, Sahra shot riders waiting in La Sombrita's little shadow. Thank you both.

Nancy Levinson, David Hajdu, and Alisa Solomon gave me the opportunity as a graduate of the Columbia Journalism School to write the deeply researched and reported article about shade for *Places Journal* that was the genesis of this book, and Josh Wallaert's skillful edits made the piece a success. Since that article was published in 2019, I have received friendly emails and been forwarded viral social media posts from readers about shade, thanks in part to Roman Mars, Chris Berube, and the staff of *99% Invisible.* To anyone who has ever sent me a story about shade or a photo of someone standing torturously in a slim shadow—you shed light, sparked inspiration, and reminded me of the broader interest and worthiness of my book.

My friends moved this book along in ways big and small. Dan Siegal, Kyle Nelson, Mary Lewine, David Knowles, Freddy Deknatel, and Nathan Uren offered close reads and feedback on chapter drafts. Lindsey Schwoeri encouraged me and helped me find an agent. I'm fortunate to have Merlin Chowkwanyun, Christopher Rice, Erik Felix, Peter Feigenbaum, Justin Smith, Amelia Hazinski, and Nicholas Simcik Arese in my life, who work in public health, urban planning, and architecture and are willing to answer casual questions, offer reality checks, and make architectural drawings and shadow studies. David's family generously hosted me during a reporting trip in Portland, and Hunter Haney and Kameron Azarm put me up in Los Angeles.

My therapist, Ian, for our productive sessions and giving me tools to manage stress and anxiety. My whole family, especially my parents, Barbara Kahn and Robert Bloch, for cheering me on, and for their understanding when I was working during holidays: Becca Kahn Bloch, Jordana Starkman, Ben Bloch, Caroline Rutherford, Ruby

Tondu, David Hankin, Sherman Chan, Mona Chang, and Amelia Chan. In New York, friends and neighbors are never far, and they kept Cybelle and Eli company when I was busy: Casey Gonzalez, Joe Siegel, Dena Yago, Sam Korman, Catie Khella, Clark Mizono, Leon Neyfakh, Alice Gregory, Noah Kardos-Fein, Amy Scheim, Nicole Hurstell Smith, Sami Rubenfeld, and Sarah Polsky. A note of appreciation for the staff of the Seward Park branch of the New York Public Library, who not only helped me with book research, but also entertained my family on Saturday mornings. And finally, my son, Eli, who was born when I was writing this book and who has shown me the world in a new light.

Notes

Introduction

ix **Cypress Park's bus riders:** Correspondence with Ash Kramer of the Greater Cypress Park Neighborhood Council, October 2018.

x **the barbershop's curmudgeonly proprietor:** Interview with Tony Cornejo, May 14, 2018. Cornejo died in 2020.

x **beneficiaries of his street shelter:** Urban planner James Rojas coined the term "Latino urbanism" to describe homespun enhancements like Cornejo's bus shelter.

x **the city's sidewalk inspectors:** Correspondence with Paul Gomez of the City of Los Angeles Department of Public Works, May and June 2018.

xi **worst heat wave in twenty-five years:** In early October 2015, downtown Los Angeles recorded triple-digit high temperatures for three days straight. See Jason Samenow, "Los Angeles Has Worst Heat Wave in 25 Years," *The Washington Post,* October 12, 2015.

xi **two to three weeks longer:** Hayley Smith, "Hotter, Drier and All-Around Different: How Climate Change Will Alter Your Life in L.A.," *Los Angeles Times,* September 9, 2024.

xii **"many, many centuries":** NASA, "Is It Too Late to Prevent Climate

Change?," n.d., accessed November 13, 2024, https://science.nasa.gov/climate-change/faq/is-it-too-late-to-prevent-climate-change/.

xii **"We all know that cities":** *CNN Newsroom,* "Parts of U.S. South and Southwest Continue to Experience Extreme Heat," CNN, July 29, 2023.

xiii **safe drinking water and clean air:** Interview with Kelly Turner, August 8, 2021.

xiii **between forty-one days of unsafe heat:** Nico Kommenda et al., "Where Dangerous Heat Is Surging," *The Washington Post,* September 5, 2023. The article includes an interactive CarbonPlan climate model that is based on real-world recordings of solar radiation in shade.

xiii **to hang out outside:** Jennifer Vanos et al., "A Physiological Approach for Assessing Human Survivability and Liveability to Heat in a Changing Climate," *Nature Communications* 14 (2023): 7653.

xiii **The Mesopotamian goddess Inanna:** Mary Shepperson, *Sunlight and Shade in the First Cities: A Sensory Archaeology of Early Iraq* (Göttingen: Vandenhoeck & Ruprecht, 2017), 49.

xiii **In Accra, Khartoum, and Kinshasa:** Lesley Lokko, "Shady Democracy," *The Architectural Review,* April 2020.

xiii **desert villages of Burkina Faso:** Iwan Baan and Francis Kéré, *Momentum of Light* (Zurich: Lars Müller, 2021).

xiv **return from Mexico and Cuba:** Mark Godfrey et al., *Francis Alÿs: A Story of Deception* (London: Tate Publishing, 2010), 102; Sam Valentine, "What Urban Designers and Landscape Architects Can Learn About Public Space From Cuba," *Common Edge,* April 8, 2019.

xiv **Marrakech, New Delhi, and Singapore:** Faisal Aljawabra and Marialena Nikolopoulou, "Thermal Comfort in Urban Spaces: A Cross-Cultural Study in the Hot Arid Climate," *International Journal of Biometeorology* 62, no. 10 (2018): 1901–09; S. Manavvi and E. Rajasekar, "Semantics of Outdoor Thermal Comfort in Religious Squares of Composite Climate: New Delhi, India," *International Journal of Biometeorology* 64, no. 2 (2020): 253–64; Wei Yang et al., "Thermal Comfort in Outdoor Urban Spaces in Singapore," *Building and Environment* 59 (January 2013): 426–35.

xiv **even temperate zones:** Marialena Nikolopoulou and Koen Steemers, "Thermal Comfort and Psychological Adaptation as a Guide for Designing Urban Spaces," *Energy and Buildings* 35, no. 1 (2003): 95–101.

xiv ***shade tree* has disappeared:** For example, the International Society of Arboriculture (ISA), the largest association of tree care professionals, was known as the International Shade Tree Conference until 1976. Correspondence with former ISA president Sharon Lilly, November 2022.

xiv **fresh as a rural spring:** In Madison, Wisconsin, scientists recorded a 10-degree swing between the most and least forested areas. Carly D. Ziter et al., "Scale-Dependent Interactions Between Tree Canopy Cover and Impervious Surfaces Reduce Daytime Urban Heat During Summer," *Proceedings of the National Academy of Sciences of the United States of America* 116, no. 15 (April 9, 2019): 7575–80.

xv **only significant predictor:** Ariane Middel et al., "Impact of Shade on Outdoor Thermal Comfort—a Seasonal Field Study in Tempe, Arizona," *International Journal of Biometeorology* 60, no. 12 (2016): 1849–61.

xv **the city itself offers:** Ariane Middel et al., "50 Grades of Shade," *Bulletin of the American Meteorological Society* 102, no. 9 (May 2021): 1–35. According to Brian Stone, Jr., building shadows reduce heat stress indices by 20 to 30 percent. See *Radical Adaptation: Transforming Cities for a Climate Changed World* (New York: Cambridge University Press, 2024): 41.

xv **the shady characters (*umbratici*):** Jeremy Hartnett, *The Roman Street: Urban Life and Society in Pompeii, Herculaneum, and Rome* (New York: Cambridge University Press, 2017), 54.

xv **"from shadowy types to truth":** John Milton, *Paradise Lost* (New York: Penguin Books, 2003).

xvi **The Greek philosopher Onesicritus:** Michael R. Dove, *Bitter Shade* (New Haven: Yale University Press, 2021), 94.

xvi **"Save Our Light":** Jake Schmidt (@TheJakeSchmidt), "You want to see something funny? Some Upper East Side NIMBYs are holding a #StopTheTower rally against the New York Blood Center project, right now. In 88° weather, in the sun, with 'Save Our Light' signs. You'll never guess where fully half the rally attendees are," Twitter (now X), May 23, 2021, https://x.com/TheJakeSchmidt/status/1396533546819497990.

xvii **warming faster than the planet:** Brian Stone, *The City and the Coming Climate: Climate Change in the Places We Live* (New York: Cambridge University Press, 2012), 86.

Chapter 1: Made in the Shade

3 **cleared this land:** Interview with U.S. Forest Service research ecologist Steve Wondzell, June 29, 2023; Wondzell, "What Will Nature Do? Shading Out Climate Change: Restoring Stream-Side Forests to Keep Streams Cool for Salmon," Corvallis Art Center, February 18, 2021, YouTube, https://www.youtube.com/watch?v=LOpwUR5w1qI.

4 **tepid water is dangerous:** Steve Wondzell et al., "What Matters Most: Are Future Stream Temperatures More Sensitive to Changing Air Temperatures, Discharge, or Riparian Vegetation?," *Journal of the American Water Resources Association* 55, no. 1 (February 2019): 116–32.

5 **99 percent of animals:** Patrice Pottier, "Young Cold-Blooded Animals Are Suffering the Most as Earth Heats Up, Research Finds," *The Conversation,* September 19, 2022.

6 **biologist Michael Sears:** Interview on March 1, 2021.

7 **ecologists call thermal refuges:** Joseph L. Ebersole et al., "Managing Climate Refugia for Freshwater Fishes Under an Expanding Human Footprint," *Frontiers in Ecology and the Environment* 18, no. 5 (June 2020): 271–80.

7 **Oregon's state environmental regulators:** State of Oregon, Department of Environmental Quality, *Fact Sheet: Pollution Limits and Water Quality Plan for the John Day River Basin* (10-ER-004), November 9, 2010.

7 **Wasco, Warm Springs, and Paiute tribes:** John Kirkland, "Shading Out Climate Change: Planting Streamside Forests to Keep Salmon Cool," *Science Findings* 228 (June 2020): 1–5.

7 **understories of rainforests:** Rhett A. Butler, "The Ground Layer of the Rainforest," WorldRainforests.com, April 19, 2019.

8 **ambient temperatures in tropical forests:** Yuta Masuda et al., "How Are Healthy, Working Populations Affected by Increasing Temperatures in the Tropics? Implications for Climate Change Adaptation Policies," *Global Environmental Change* 56 (2019): 29-40.

8 **clouds in the atmosphere:** George A. Ban-Weiss et al., "Climate Forcing and Response to Idealized Changes in Surface Latent and Sensible Heat," *Environmental Research Letters* 6, no. 3 (2011): 034032.

9 **biogeographer Greg Barron-Gafford:** Interviews in June and July 2021.

9 **African and Middle Eastern farmers:** Wolfram Achtnich, "Shading in Plant Production," in *Shadow in the Desert,* eds. Andrea Bienhaus and Ewald Bubner (Stuttgart: Heinrich Fink, 1972), 19–27.

10 **Brad Heins, a University of Minnesota:** Interview on March 31, 2023.

10 **Temple Grandin has toured feedyards:** Interview on April 17, 2023.

10 **desert rodents avoid overheating:** E. A. Riddell et al., "Exposure to Climate Change Drives Stability or Collapse of Desert Mammal and Bird Communities," *Science* 371, no. 6529 (February 5, 2021): 633–36.

10 **northern face of a tree:** All shadow projections in this book are in the Northern Hemisphere, unless noted.

10 **their avian counterparts nest:** Interview with University of New Mexico biology professor Blair Wolf, June 8, 2021.

11 **tough life on the Senegalese savanna:** Carl Zimmer, "Hints of Human Evolution in Chimpanzees That Endure a Savanna's Heat," *The New York Times,* April 27, 2018.

11 **sophisticated visual system:** Interview with Lund University professor emeritus Dan-Eric Nilsson, December 28, 2022.

11 **infrared waves:** Paweł Sowa et al., "Optical Radiation in Modern Medicine," *Advances in Dermatology and Allergology* 30, no. 4 (August 2013): 246–51; Luke Horton et al., "The Effects of Infrared Radiation on the Human Skin," *Photodermatology, Photoimmunology & Photomedicine* 39, no. 6 (November 2023): 549–55.

12 **molecular thermometers turn on:** Ken Parsons, *Human Thermal En-*

vironments: The Effects of Hot, Moderate, and Cold Environments on Human Health, Comfort, and Performance, 3rd ed. (Boca Raton, Fla.: CRC Press, 2002), 70.

12 **according to Zachary Schlader:** Interview on August 11, 2022.

13 **Seeking shade spares us:** Zachary Schlader et al., "Characteristics of the Control of Human Thermoregulatory Behavior," *Physiology & Behavior* 98, no. 5 (December 7, 2009): 557–62.

13 **destined for the insula:** Andreas D. Flouris, "Functional Architecture of Behavioural Thermoregulation," *European Journal of Applied Physiology* 111 (January 2011): 1–8.

14 **the "forest edge" effect:** Robert Geddes, *The Forest Edge* (New York: St. Martin's, 1982).

14 **thermal-neural hardware:** Sally T. Sauter et al., *Issue Paper 1: Salmonid Behavior and Water Temperature* (Washington, D.C.: U.S. Environmental Protection Agency, 2001); Daniel Mota-Rojas et al., "Physiological and Behavioral Mechanisms of Thermoregulation in Mammals," *Animals* 11, no. 6 (June 10, 2021): 1733.

14 **The Canadian neuroscientist Michel Cabanac:** Michel Cabanac, "Physiological Role of Pleasure," *Science* 173, no. 4002 (September 17, 1971): 1103–07; Richard de Dear, "Revisiting an Old Hypothesis of Human Thermal Perception: Alliesthesia," *Building Research & Information* 39, no. 2 (2011): 108–17.

14 **six or seven times as many cold thermoreceptors:** Parsons, *Human Thermal Environments,* 90.

15 **marvel at the speed:** Correspondence with Ken Parsons, September 2024.

15 **euphoric sensation an "overshoot":** Yuliya Dzyuban et al., "Evidence of Alliesthesia During a Neighborhood Thermal Walk in a Hot and Dry City," *Science of the Total Environment* 834 (August 15, 2022): 155294.

15 **only skin-deep:** Flouris, "Functional Architecture of Behavioural Thermoregulation."

15 **Discomfort is a thermal alert:** Michel Cabanac, "Pleasure: The Common Currency," *Journal of Theoretical Biology* 155, no. 2 (March 21, 1992): 173–200.

Chapter 2: Shady Lanes

16 **In the Mesopotamian flats:** Mary Shepperson, *Sunlight and Shade in the First Cities: A Sensory Archaeology of Early Iraq* (Göttingen: Vandenhoeck & Ruprecht, 2017).

17 **"It's one of the reasons":** Interview with Mary Shepperson, January 2, 2021.

18 **In the medina of Fez:** Erik Johansson, "Influence of Urban Geometry on Outdoor Thermal Comfort in a Hot Dry Climate: A Study in Fez, Morocco," *Building and Environment* 41, no. 10 (2006): 1326–38.

19 **compact urban form:** Shepperson, *Sunlight and Shade,* 89–90; Hassan Fathy, *Natural Energy and Vernacular Architecture: Principles and Examples with Reference to Hot Arid Climates* (Chicago: University of Chicago Press, 1986), 64.

19 **"Let them build":** Shepperson, *Sunlight and Shade,* 56.

20 **"shine like the day":** Shepperson, *Sunlight and Shade,* 101.

20 **shading them with parasols:** Oscar White Muscarella, "Parasols in the Ancient Near East," *Source: Notes in the History of Art* 18, no. 2 (Winter 1999): 1–7.

20 **"It's an improbable city":** John Berger, *The Red Tenda of Bologna* (London: Penguin Classics, 2018), 19.

21 **originated in ancient Greece:** J. J. Coulton, *The Architectural Development of the Greek Stoa* (Oxford: Oxford University Press, 1976); Spiro Kostof, *A History of Architecture: Settings and Rituals,* 2nd ed. (New York: Oxford University Press, 1995), 143–45.

21 ***peripatoi* (shady walks):** John Patrick Lynch, *Aristotle's School: A Study of a Greek Educational Institution* (Berkeley: University of California Press, 1972), 72.

21 **From the Greeks:** Coulton, *Greek Stoa,* 178; Fikret Yegül and Diane Favro, *Roman Architecture and Urbanism: From the Origins to Late Antiquity* (Cambridge: Cambridge University Press, 2018).

21 **A typical covered sidewalk:** Interview with Diane Favro, December 15, 2022.

22 **"And people hated it":** Favro interview.

22 **the porticoes formed a shadow network:** Rodolfo Lanciani, *The Ruins and Excavations of Ancient Rome: A Companion Book for Students*

and Travelers (Boston and New York: Houghton Mifflin, 1897), 445–46.

22 **One could effectively stroll:** Timothy O'Sullivan, "The Mind in Motion: The Cultural Significance of Walking in the Roman World" (PhD diss., Harvard University, 2003), 9.

22 **point to game boards:** Benjamin Miles Crowther, "Life on the Streets: Architecture and Community Along the Colonnaded Streets of the Roman Empire" (PhD diss., University of Texas at Austin, 2019).

23 **The more desperate members:** Jeremy Hartnett, *The Roman Street: Urban Life and Society in Pompeii, Herculaneum, and Rome* (New York: Cambridge University Press, 2017).

23 **Hellenistic pleasure walks:** Elizabeth Macaulay-Lewis, "The City in Motion: Walking for Transport and Leisure in the City of Rome," in *Rome, Ostia, Pompeii: Movement and Space,* eds. Ray Laurence and David J. Newsome (New York: Oxford University Press, 2011), 262–89.

23 **"association is not altered":** Libanius, "Oration in Praise of Antioch (Oration XI)," trans. Glanville Downey, *Proceedings of the American Philosophical Society* 103, no. 5 (October 15, 1959): 652–86.

23 **hard to imagine:** Warwick Ball, *Rome in the East: The Transformation of an Empire* (London: Routledge, 2001), 262–63, 269.

24 **In the thirteenth century:** Samuel Packard, "The Porticoes of Bologna," *Landscape* 27, no. 1 (January 1983): 19–29.

24 **capped the heights:** The concern was not structural instability but the launching of projectiles. See Naomi Miller, *Renaissance Bologna: A Study in Architectural Form and Content* (New York: Peter Lang, 1989), 30–31.

24 **Under papal rule:** Kathleen Olive, "Bologna's Porticoes," Istituto Italiano di Cultura, Sydney, September 29, 2021, YouTube, https://www.youtube.com/watch?v=Ztmbtd6tQuM.

24 **compulsory by statute:** Although other European cities boast arcaded sidewalks, such as Paris, Bern, and Padua, Bologna is the only one to legally compel them, according to Federica Legnani, a city architect who oversees Bologna's "regeneration" of historical urban landscapes and porticoes (interview on June 23, 2021).

24 **porticoes became outdoor workshops:** Francesca Bocchi and Rosa Smurra, "Bologna and Its Porticoes: A Thousand Years' Pursuit of the 'Common Good,'" *Quart* 3 (2020): 87–104.

24 **"altruism turned architecture":** Bernard Rudofsky, *Architecture Without Architects* (Garden City, N.Y.: Doubleday, 1964).

25 **"Seville is a city of shadows":** V. S. Pritchett, *The Spanish Temper* (New York: Knopf, 1954), 175.

27 **they also appreciated awnings:** Favro interview.

27 **introduced by the Moors:** Interview with Álvaro Pimentel Siles, July 27, 2022. See also Simon Schleider, "Adaptive Toldo Systems" (March thesis, Massachusetts Institute of Technology, 2009), 10.

27 **discarded from ships:** Pedro Ybarra Bores, "¿Desde cuándo se ponen los toldos en el Centro de Sevilla?," *ABC de Sevilla,* July 9, 2021.

28 **According to writing:** The Spanish historian José Gestoso relayed this description of the Corpus Christi festival in Seville. See Bores, "¿Desde cuándo?," *ABC de Sevilla.*

28 **Daniel González Rojas, a scruffy:** Interview on July 25, 2022. González Rojas left the city council that year.

29 **Climate activists had resorted:** Antonio Morente, "Sevilla por el Clima sale bajo palio para reclamar toldos que den sombra en los puentes," *elDiario.es,* June 29, 2021.

29 **Álvaro Pimentel Siles, a conservative:** Siles interview.

30 ***verano sin sombra:*** Juan Parejo, "Sevilla se queda sin toldos en las calles en un verano de mucho calor," *Diario de Sevilla,* July 1, 2021.

31 **"considerable convenience":** Axel I. Mundigo and Dora P. Crouch, "The City Planning Ordinances of the Laws of the Indies Revisited. Part I: Their Philosophy and Implications," *The Town Planning Review* 48, no. 3 (July 1977): 247–68; Dora P. Crouch and Axel I. Mundigo, "The City Planning Ordinances of the Laws of the Indies Revisited. Part II: Three American Cities," *The Town Planning Review* 48, no. 4 (October 1977): 397–418.

31 **Lima and downtown Los Angeles:** Jeffrey Aronin, *Climate & Architecture* (New York: Reinhold, 1953), 10; Ralph Knowles, *Sun Rhythm Form* (Cambridge, Mass.: MIT Press, 1981), 20–21.

31 **dusty wood boardwalks:** Correspondence with University of New Mexico emeritus professors Virginia Scharff and Chris Wilson, October 2022.

32 **stunning cliff dwellings:** Ralph Knowles, *Energy and Form: An Ecological Approach to Urban Growth* (Cambridge, Mass.: MIT Press, 1980), 20–26.

32 **The Hohokam of Arizona:** Stan Cox, *Losing Our Cool: Uncomfortable Truths About Our Air-Conditioned World (and Finding New Ways to Get Through Summer)* (New York: The New Press, 2010), 1–3.

32 **most influential colonists, the British:** John E. Crowley, *The Invention of Comfort: Sensibilities and Design in Early Modern Britain and Early America* (Baltimore: The Johns Hopkins University Press, 2000), 87, 230–59.

33 **William Penn came to the New World:** John W. Reps, *The Making of Urban America: A History of City Planning in the United States* (Princeton, N.J.: Princeton University Press, 1965), 157–74.

33 **"greene country towne":** Samuel Fitch Hotchkin, *Penn's Greene Country Towne: Pen and Pencil Sketches of Early Philadelphia and its Prominent Characters* (Philadelphia: Ferris & Leach, 1903).

33 **American cities like Savannah:** Stanford University Libraries, Barry Lawrence Ruderman Map Collection, "A Portraiture of the City of Philadelphia in the Province of Pennsylvania in America. [with] A Letter from William Penn Proprietary and Governour of Pennsylvania in America, to the Committee of the Free Society of Traders," n.d., accessed November 11, 2024, purl.stanford.edu/jj125yt6481.

33 **"light and airy":** Reps, *Making of Urban America,* 246.

34 **realm of savages:** Keith Thomas, *Man and the Natural World: Changing Attitudes in England, 1500–1800* (London: Penguin Books, 1984), 194.

34 **spared on farms:** Thomas Campanella, *Republic of Shade: New England and the American Elm* (New Haven: Yale University Press, 2003), 25.

34 **scarce on the streets:** Anne Beamish, "A Garden in the Street: The Introduction of Street Trees in Boston and New York," *Studies in the History of Gardens & Designed Landscapes* 38, no. 1 (2018): 38–56.

35 **breakthrough moment occurred:** Beamish, "Garden in the Street."

35 **Boston Common was totally denuded:** Campanella, *Republic of Shade,* 70.

35 **squares of Savannah:** Jefferson Hall, "The Squares: The True History of Savannah's Tree History," *Fact-Checking Savannah's History,* February 8, 2020, https://savannahhistory.home.blog/2020/02/08/the-squares-the-true-history-of-savannahs-tree-history.

35 **The architectural historian Michael Webb:** Michael Webb, *The City Square* (London: Thames & Hudson, 1990), 116–18.

35 **third force was yellow fever:** Beamish, "Garden in the Street."

35 **cities would be deserted:** Kenneth T. Jackson, *Crabgrass Frontier: The Suburbanization of the United States* (New York: Oxford University Press, 1985), 58.

35 **caused by miasma:** William B. Meyer, "Urban Heat Island and Urban Health: Early American Perspectives," *Professional Geographer* 43, no. 1 (1991): 38–48.

36 **"Another thing to be observed":** Noah Webster, *A Brief History of Epidemic and Pestilential Diseases* (Hartford, Conn.: Hudson and Goodwin, 1799).

36 **Charles Caldwell, like Webster:** Charles Caldwell, *Medical and Physical Memoirs* (Philadelphia: Thomas and William Bradford, 1801).

37 **Doctors came to urge:** Beamish, "Garden in the Street."

37 **European city planners:** Henry W. Lawrence, *City Trees: A Historical Geography from the Renaissance Through the Nineteenth Century* (Charlottesville: University of Virginia Press, 2006).

37 **"*water, air, and shade*":** Claude Philibert Barthelot Rambuteau, *Memoirs of the Comte de Rambuteau: Edited by His Grandson,* trans. J.C. Brogan (New York: Putnam, 1908), 208.

38 **"lungs of the city":** Karen R. Jones, "Green Lungs and Green Liberty: The Modern City Park and Public Health in an Urban Metabolic Landscape," *Social History of Medicine* 35, no. 4 (November 2022): 1200–22.

38 **The park's designer, Frederick Law Olmsted:** Charles E. Beveridge, *Mount Royal in the Works of Frederick Law Olmsted* (Montréal: Ville de Montréal, 2009), 43.

38 **One such émigré was Julius Sterling Morton:** Chris Helzer, "A Prairie Ecologist's Perspective on Arbor Day," *The Prairie Ecologist,* April 26, 2013, https://prairieecologist.com/2013/04/26/a-prairie-ecologists

-perspective-on-arbor-day; Chris Helzer, "The Darker Side of Tree Planting in the Great Plains," *The Prairie Ecologist,* April 26, 2021, https://prairieecologist.com/2021/04/26/the-darker-side-of-tree-planting-in-the-great-plains.

38 **"battle against the treeless prairies":** Byron Anderson, "Biographical Portrait: Julius Sterling Morton," *Forest History Today* (Fall 2000): 31–33.

38 **planting the arcaded plazas:** Spiro Kostof, *The City Assembled: The Elements of Urban Form Through History* (London: Thames & Hudson, 1992), 165. In smaller frontier towns, settlers from New England transformed Spanish plazas into green courthouse squares. See David Handlin, *The American Home: Architecture and Society, 1815–1915* (Boston: Little, Brown, 1979), 112–13.

39 **health justification for shade:** John M. Harris, Jr., "Manhattan's Street Trees: An Unfinished Public Health Story," *American Journal of Public Health* 115, no. 1 (January 2025): 66–74.

39 **Olmsted urged cities:** Frederick Law Olmsted, "Public Parks and the Enlargement of Towns," February 25, 1870, in *The Papers of Frederick Law Olmsted: Supplementary Series, Volume I: Writings on Public Parks, Parkways, and Park Systems,* ed. Charles E. Beveridge and Carolyn F. Hoffman (Baltimore: The Johns Hopkins University Press, 1997), 171–201.

39 **The scant street trees:** Max Page, "'Uses of the Axe': Towards a Treeless New York," *American Studies* 40, no. 1 (Spring 1999): 41–64.

39 **Stephen Smith, a physician:** Harris, "Manhattan's Street Trees."

39 **"The day would not seem":** Stephen Smith, "Vegetation: A Remedy for the Summer Heat of Cities," *Appletons' Popular Science Monthly,* ed. William Jay Youmans, vol. 54 (February 1899): 441.

Chapter 3: Climate Control

41 **July 17, 1902:** James Barron, "More Than Pipe Dream, It Was the Idea That Led to Air-Conditioning," *The New York Times,* July 17, 2012.

42 **workers more productive:** Gail Cooper, *Air-Conditioning America: Engineers and the Controlled Environment, 1900–1960* (Baltimore: The Johns Hopkins University Press, 1998), 55–58.

42 **New York Stock Exchange:** Eric Dean Wilson, *After Cooling: On Freon, Global Warming, and the Terrible Cost of Comfort* (New York: Simon & Schuster, 2021), 46–49.

42 **"The boys don't even know":** Marsha Ackermann, *Cool Comfort: America's Romance with Air-Conditioning* (Washington, D.C.: Smithsonian Books), 66.

42 **"comfort cooling":** Raymond Arsenault, "The End of the Long Hot Summer: The Air Conditioner and Southern Culture," *The Journal of Southern History* 50, no. 4 (November 1984): 597–628.

43 **88 percent today:** U.S. Energy Information Administration 2020 Residential Energy Consumption Survey.

43 **desire to live outdoors:** Andrea Vesentini, *Indoor America: The Interior Landscape of Postwar Suburbia* (Charlottesville: University of Virginia Press, 2018).

43 **AC saved lives:** Alan Barreca et al., "Adapting to Climate Change: The Remarkable Decline in the U.S. Temperature-Mortality Relationship over the 20th Century," National Bureau of Economic Research Working Paper 18692, January 2013.

43 **makes glassy buildings viable:** Stephen Eskilson, *The Age of Glass* (New York: Bloomsbury Academic, 2018), 169.

44 **T-shaped and U-shaped:** Patrick Sisson, "How Air Conditioning Shaped Modern Architecture—and Changed Our Climate," *Curbed,* May 9, 2017.

44 **retractable canvas awning:** Adam Brodheim, "'Many Awnings for Manifold Uses': The Rise and Prevalence of Window Awnings in Late 19th and Early 20th Century American Architecture," Columbia University Graduate School of Architecture, Planning and Preservation, 2021.

44 **"when real summer weather":** Robert Khederian, "How Houses Were Cooled Before Air Conditioning," *Curbed,* July 20, 2017.

45 **awning service to tenants:** Salvatore Basile, *Cool: How Air Conditioning Changed Everything* (New York: Fordham University Press, 2014), 173.

45 **iconic Flatiron Building:** Brodheim, "'Many Awnings for Manifold Uses.'"

45 **Florida state capitol:** Laura A. Muckenfuss and Charles E. Fisher, "Windows Number 7: Window Awnings," *Preservation Tech Notes* (Washington, D.C.: National Park Service, 1984).

45 **Seattle and New Orleans and Barcelona and Paris:** Seattle architect Michael Eliason posted photos of window awnings, taken between 1908 and 1912, on Twitter (@holz_bau) on August 9, 2022.

45 **commercial storefronts nailed:** Chad Randl, *Preservation Briefs 44: The Use of Awnings on Historic Buildings: Repair, Replacement, and New Design* (Washington, D.C.: National Park Service, 2005).

45 **New York, Philadelphia, or Baltimore:** Randl, *Use of Awnings on Historic Buildings;* Bernard Rudofsky, *Streets for People: A Primer for Americans* (Garden City, N.Y.: Doubleday, 1969).

45 **Bowery was a sidewalk:** Catherine Hoover Voorsanger and John K. Howat, eds., *Art and the Empire City: New York, 1825–1861* (New York: The Metropolitan Museum of Art, 2000), 20, 235.

45 **permanent canopies of sheet metal:** In the 1910s and 1920s, *The American City,* an influential urban planning magazine, recorded the removal of wood and sheet metal awnings from sidewalks in Poughkeepsie, New York; Allentown, Pennsylvania; Frederick, Maryland; Lynchburg, Virginia; Clarksburg, West Virginia; Elberton, Georgia; Tallahassee, Florida; and Monterey and Stockton, California, among other places.

46 **popular in New Orleans:** Interview with Tulane University geographer Richard Campanella, October 18, 2022. See also Richard Campanella, "Where Have All the Awnings Gone? The Shady History of New Orleans' Historic Sidewalk Feature," *The Times-Picayune,* August 8, 2024.

46 **Urban congestion:** Anastasia Loukaitou-Sideris and Renia Ehrenfeucht, *Sidewalks: Conflict and Negotiation over Public Space* (Cambridge, Mass.: MIT Press, 2009), 21–27.

46 **danger of daytime darkness:** I borrow this phrase from David Gissen, "The Dark Day Returns," *Art in America* 108, no. 5 (May 2020): 34–39.

46 **Reformers like Robert DeForest:** Daniel Freund, *American Sunshine:*

Diseases of Darkness and the Quest for Natural Light (Chicago: University of Chicago Press, 2012), 11–12.

46 **shadows of new skyscrapers:** Robert Fogelson, *Downtown: Its Rise and Fall, 1880–1950* (New Haven: Yale University Press, 2001), 125–26.

47 **Boston architect named William Atkinson:** William Atkinson, *The Orientation of Buildings or Planning for Sunlight* (New York: John Wiley & Sons, 1912).

47 **build bigger roads:** Michael Southworth and Eran Ben-Joseph, *Streets and the Shaping of Towns and Cities* (Washington, D.C.: Island Press, 2009), 5–6, 12.

47 **huge Viennese Ringstrasse:** David Gissen, *The Architecture of Disability: Buildings, Cities, and Landscapes Beyond Access* (Minneapolis: University of Minnesota Press, 2022), 164–65.

47 **unsanctioned "encroachments":** The City Plan Commission, Newark, New Jersey, *Comprehensive Plan of Newark* (Newark: H. Murphy, 1915): 102–04. See also Peter Norton, *Fighting Traffic: The Dawn of the American Motor Age in the American City* (Cambridge, Mass.: MIT Press, 2011), 134–40.

47 **"these ugly affairs":** Harland Bartholomew, *Plans for the Development of a System of Major Streets, Evansville, Indiana* (Evansville: City Plan Commission, 1925), 12; Harland Bartholomew, *The Lansing Plan: A Comprehensive City Plan Report for Lansing, Michigan* (Lansing: R. Smith, 1922), 52-53.

47 **"unsightly" sheds and awnings:** M. V. Fuller, "The Rejuvenation of Poughkeepsie," *The American City* 4, no. 1 (January 1911): 3; G. D. Theleen, "Main Street, Before and After," *The American City* 29, no. 2 (August 1923): 113; J. Herbert Kohler, "Up-to-Date Business Streets," *The American City* 20, no. 1 (January 1919): 56–57.

48 **building heights commission:** Jason M. Barr, "Revisiting 1916 (Part 1): The History of New York City's First Zoning Resolution," Building the Skyline, March 27, 2019, https://buildingtheskyline.org/revisiting-1916-i.

48 **construction of the Equitable Building:** Nolan Gray, "Light and Air,

Sound and Fury; or, Was the Equitable Life Building Panic Only About Shadows?," Market Urbanism, September 6, 2018, https://marketurbanism.com/2018/09/06/nyc-zoning-1916-equitable-life-building.

48 **Bassett later recalled:** Ruth Knack et al., "The Real Story Behind the Standard Planning and Zoning Acts of the 1920s," *Land Use Law & Zoning Digest* 48, no. 2 (February 1996): 3–9.

48 **around 220 municipalities:** Freund, *American Sunshine,* 23–24.

49 **"common knowledge":** Freund, *American Sunshine,* 24.

49 **In Baltimore, planners:** Emily Talen, *City Rules: How Regulations Affect Urban Form* (Washington, D.C.: Island Press, 2011), 146–47.

49 **single-family houses and setbacks:** Talen, *City Rules,* 64; Southworth and Ben-Joseph, *Streets and the Shaping of Towns and Cities,* 87, 96.

49 **Yard Ordinance:** Andrew Whittemore, "How the Federal Government Zoned America: The Federal Housing Administration and Zoning," *Journal of Urban History* 39, no. 4 (2012): 620–42.

49 **Swiss-French architect Le Corbusier:** Le Corbusier et al., "Glass, the Fundamental Material of Modern Architecture," *West 86th: A Journal of Decorative Arts, Design History, and Material Culture* 19, no. 2 (Fall–Winter 2012): 282–308.

50 **"thirty years in the dark":** Margaret Campbell, "What Tuberculosis Did for Modernism: The Influence of a Curative Environment on Modernist Design and Architecture," *Medical History* 49, no. 4 (October 2005): 463–88.

50 **United Nations Secretariat:** David Arnold, "Air Conditioning in Office Buildings After World War II," *ASHRAE Journal* 41, no. 7 (July 1999): 33–41.

50 **"Something should be said":** David Huber, "Hothouses: Architecture and the Onslaught of Heat Fatigue," *PIN-UP* 28 (Spring-Summer 2020).

50 **Ludwig Mies van der Rohe:** Alex Beam, *Broken Glass: Mies van der Rohe, Edith Farnsworth, and the Fight over a Modernist Masterpiece* (New York: Random House, 2020).

51 **"It's up to the engineers":** Barnabas Calder and Florian Urban, "It Is

Time No Longer to Praise the Seagram Building, but to Bury It," *Architects Journal,* November 10, 2022.

51 **only air-conditioning makes it work:** Cooper, *Air-Conditioning America,* 161–62.

51 **adapted their Victorian cottages:** Interview with Demion Clinco of the Tucson Historic Preservation Foundation, July 28, 2021.

51 **Seminoles built open-walled shelters:** Amos Rapoport, *House Form and Culture* (Englewood Cliffs, N.J.: Prentice-Hall, 1969), 94.

52 **after World War II:** Adam Rome, *The Bulldozer in the Countryside: Suburban Sprawl and the Rise of American Environmentalism* (New York: Cambridge University Press, 2001).

52 **indoor climate-makers:** "Air Conditioning: Now for Small Homes," *Newsweek,* September 8, 1952, 73.

52 **"building a Florida room":** Arsenault, "End of the Long Hot Summer," 624.

53 **rough in for AC:** Cooper, *Air-Conditioning America,* 157.

53 **trusty old shade tree:** Jule R. von Sternberg, "The Economics of Trees," *House & Home,* April 1953, 130–36.

53 **"house in a treeless tract":** A. M. Watkins, "Five Top Priorities for Designing," *House & Home,* August 1953, 100.

53 **That made "air cooling":** "Round Table Report: Part 1: The Conditioned House," *House & Home,* April 1961, 107.

54 **"the soul can rest":** bell hooks, *Belonging: A Culture of Place* (New York: Routledge, 2009), 143–52.

54 **Frank Trippett anticipated:** Frank Trippett, "The Great American Cooling Machine," *Time,* August 13, 1979.

54 **the threat of long-term scarcity:** Rome, *Bulldozer in the Countryside,* 45–51; Daniel Barber, *Modern Architecture and Climate: Design Before Air Conditioning* (Princeton, N.J.: Princeton University Press, 2020).

55 **two decades' worth of weather data:** Barber, *Modern Architecture,* 171–72.

55 **"doubly important now":** Housing and Home Finance Agency, *Climate and Architecture: Selected References* (Washington, D.C.: March 1951), i.

55 **Joseph Orendorff, the agency's research director:** "Building Indus-

try Warned to Spend More on Research or Face Rising Gov't Control," *House & Home,* June 1952, 55.

55 **"indispensable to the design":** Housing and Home Finance Agency, *Application of Climatic Data to House Design* (Washington, D.C.: January 1954), ii.

55 **Victor and Aladar Olgyay:** Barber tells the definitive story of the brothers in *Modern Architecture,* 199–245.

55 **pinpoint man's "comfort zone":** Eric Teitelbaum et al., "Highway to the Comfort Zone: History of the Psychrometric Chart," *Buildings* 13, no. 3 (2023): 797.

57 **designing with the climate:** Housing and Home Finance Agency, *Application of Climatic Data,* 40–90.

58 **create that shade:** Aladar Olgyay and Victor Olgyay, *Solar Control and Shading Devices* (Princeton, N.J.: Princeton University Press, 1957).

58 **solar hardware was critically functional:** According to the late architect Andrés Mignucci Giannoni, shading devices were written into Puerto Rican and Venezuelan health and building codes earlier in the twentieth century. See "Inhabiting Shadows: Observations on the Tropics as Place," *Places* 12, no. 3 (1999): 38–42.

58 ***House Beautiful* began to explore:** Monica Penick, *Tastemaker: Elizabeth Gordon,* House Beautiful, *and the Postwar American Home* (New Haven: Yale University Press, 2017).

60 **shelter magazines:** Anthony Paletta, "The Lost World of the Middlebrow Tastemaker," *The American Conservative,* June 8, 2018.

60 **Victory Houses:** Barber, *Modern Architecture,* 173.

61 **"harness on the sun":** Henry Wright, "How to Put a Harness on the Sun," *House Beautiful* 91, no. 10 (October 1949): 158–61, 220–22.

62 **"tolerable discomfort":** Barber, *Modern Architecture,* 212.

62 **mass-designed to conform:** Rome, *Bulldozer in the Countryside,* 61–62, 135.

62 **subvert the norms:** Ackermann, *Cool Comfort,* 118.

62 **Glass got better:** Eskilson, *Age of Glass,* 184.

63 **Seattle architect Michael Eliason:** Interview on July 9, 2021.

63 **New Yorkers escaped stifling apartments:** Arthur Miller, "Before Air-Conditioning," *The New Yorker,* June 14, 1998.

63 **People in Boston, Chicago:** Eric Klinenberg, *Heat Wave: A Social Autopsy of Disaster in Chicago* (Chicago: University of Chicago Press, 2002), 57.

63 **In 2018, Rowan Moore:** Rowan Moore, "An Inversion of Nature: How Air Conditioning Created the Modern City," *The Guardian,* August 14, 2018.

64 **hours dwindle as budgets shrink:** Maggie Gordon, "Houston Public Library's New Director Cynthia Wilson Talks Fixing Morale, Other Big Changes," *Houston Landing,* April 24, 2024.

64 **security officers bother visitors:** J. Brian Phillips, "Houston: Subterranean Treasures," *Reason,* June 1988.

64 **The tunnels exclude people:** Bill Murphy, "Downtown Houston Tunnels Unkind to Wheelchair Users," *Houston Chronicle,* August 18, 2008.

64 **"the air-conditioned nightmare":** Henry Miller, *The Air-Conditioned Nightmare* (New York: New Directions, 1945).

64 **our reliance on AC:** Negin Nazarian et al., "Integrated Assessment of Urban Overheating Impacts on Human Life," *Earth's Future* 10, no. 8 (August 23, 2022).

65 ***Losing Our Cool:*** Stan Cox, *Losing Our Cool: Uncomfortable Truths About Our Air-Conditioned World (and Finding New Ways to Get Through Summer)* (New York: The New Press, 2010), x–xi, 6–7, 110–11, 118–19.

65 **the " 'oh shit' moments":** Stephen Buranyi, "The Air Conditioning Trap: How Cold Air Is Heating the World," *The Guardian,* August 29, 2019.

65 **huge one in the U.S. Northeast:** G. Brooke Anderson and Michelle L. Bell, "Lights Out: Impact of the August 2003 Power Outage on Mortality in New York, NY," *Epidemiology* 23, no. 2 (March 2012): 189–93.

66 **risked human health:** Shao Lin et al., "Health Impact in New York City During the Northeastern Blackout of 2003," *Public Health Reports* 126, no. 3 (May–June 2011): 384–93.

66 **blacked out during a heat wave:** Brian Stone, Jr., et al., "How Blackouts During Heat Waves Amplify Mortality and Morbidity Risk," *Environmental Science & Technology* 57, no. 22 (June 6, 2023): 8245–55.

66 **"You do remember":** Interview with Brian Stone, Jr., June 19, 2020.

66 **When Phoenix officials:** Christopher Flavelle, "A New, Deadly Risk for Cities in Summer: Power Failures During Heat Waves," *The New York Times,* May 3, 2021.

66 **Phoenix didn't have:** Stone interview, February 3, 2022.

66 **"greater climate resilience":** Neel Dhanesha, "Don't Be Too Chill About Your Air Conditioning Dependency," *Heatmap News,* May 9, 2023.

67 **low-ceilinged subway stations:** Mira Wassef, "NYC Subway Stations Sizzle in the Summer. Here's Why," PIX11, July 21, 2022.

67 **used infrared cameras:** Desmond Ng and Tang Hui Huan, "Why Singapore Is Heating Up Twice as Fast as the Rest of the World," CNA, January 13, 2019.

67 **Computer simulations from Paris and Phoenix:** Francisco Salamanca Palou et al., "Anthropogenic Heating of the Urban Environment Due to Air Conditioning," *Journal of Geophysical Research: Atmospheres* 119 (2014): 5949–65.

67 **someone else's thermal garbage:** Mike Davis, "The Radical Politics of Shade," *Capitalism Nature Socialism* 8, no. 3 (September 1997): 35–39.

67 **1,300 times more effective:** Wilson, *After Cooling,* 281.

67 **federal scientists estimate:** National Renewable Energy Laboratory, "News Release: Scientists Show Large Impact of Controlling Humidity on Greenhouse Gas Emissions," March 14, 2022.

68 **how much energy:** U.S. Energy Information Administration, "Use of Energy Explained: Energy Use in Homes," December 18, 2023 (updated), https://www.eia.gov/energyexplained/use-of-energy/electricity-use-in-homes.php.

68 **10 percent of all greenhouse gas emissions:** Stan Cox, "With Air-Conditioning, Have We Passed the Point of No Return?," *The Nation,* June 17, 2024.

68 **By 2050:** International Energy Agency, "Air Conditioning Use Emerges as One of the Key Drivers of Global Electricity-Demand Growth," IEA.org, May 15, 2018.

68 **If the past is prologue:** Lindsey M. Roberts, "Air Conditioning: A Boon and a Burden," *Architect Magazine,* July 17, 2012.

Chapter 4: Surviving the Sun

71 **Maria Isabel Vasquez Jimenez:** This story is based on interviews with Marc Grossman on March 16, 2021, and Giev Kashkooli on May 13, 2021, both of the United Farm Workers (UFW); prepared remarks of former UFW president Arturo S. Rodriguez at the funeral service on May 28, 2008; and contemporaneous news articles including Susan Ferriss, "Teen Farmworker's Death, Probed as Heat-Related, Stirs Outcry," *The Sacramento Bee,* May 29, 2008; Susan Ferriss, "Coroner Confirms Teen Farmworker Died of Heat Stroke," *The Sacramento Bee,* June 18, 2008; Susan Ferriss, "Dying to Work: Dozen Farm Deaths in California Since May Linked to High Heat," *The Sacramento Bee,* August 21, 2008; Gosia Wozniacka, "Farm Supervisors Take Plea Deal in CA Heat Death," Associated Press, March 9, 2011; and Jennie Rodriguez, "No Jail for Farm Labor Officials," *The Record* (Stockton), March 10, 2011.

74 ***se quemó:*** Sarah Bronwen Horton, *They Leave Their Kidneys in the Fields: Illness, Injury, and Illegality Among U.S. Farmworkers* (Oakland: University of California Press, 2016), 32.

74 **"This land gives us":** Jennie Rodriguez, "Grief and Anger," *The Record* (Stockton), May 29, 2008.

74 **Marc Schenker agrees:** Interview on May 12, 2020.

74 **no deadlier place to work:** Diane M. Gubernot et al., "Characterizing Occupational Heat-Related Mortality in the United States, 2000–2010: An Analysis Using the Census of Fatal Occupational Injuries Database," *American Journal of Industrial Medicine* 58, no. 2 (February 2015): 203–11; "Heat-Related Deaths Among Crop Workers—United States, 1992–2006," *MMWR Weekly* 57, no. 24 (June 20, 2008): 649–53.

74 **"climate canaries":** Cora Roelofs and David Wegman, "Workers: The Climate Canaries," *American Journal of Public Health* 104, no. 10 (October 2014): 1799–801.

75 **disparaged deserts:** Baron de Montesquieu, *The Spirit of the Laws,* trans. Thomas Nugent (New York: Hafner, 1949), 240.

75 **Japanese researchers strapped sensors:** Hidenori Otani et al., "Time-of-Day Effects of Exposure to Solar Radiation on Thermoregulation During Outdoor Exercise in the Heat," *Chronobiology International* 34, no. 9 (September 2017): 1224–38.

75 **exposed to simulated sunshine:** Leonidas G. Ioannou et al., "The Impacts of Sun Exposure on Worker Physiology and Cognition: Multi-Country Evidence and Interventions," *International Journal of Environmental Research and Public Health* 18 (2021): 7698.

75 **backs of their necks:** Jacob F. Piil et al., "Direct Exposure of the Head to Solar Heat Radiation Impairs Motor-Cognitive Performance," *Scientific Reports* 10 (2020): 7812.

76 **Leonidas Ioannou, a physiologist:** Interview on September 16, 2024.

76 **two groups of workers:** Ioannou et al., "Impacts of Sun Exposure."

77 **According to Robert Brown:** In winter, when air temperature is much lower than skin temperature, the convective heat loss outweighs the amount of heat gained from the sun. Interview on November 18, 2024.

78 **"For some of us":** Interview with Jennifer Longdon, August 10, 2021.

78 **As we age:** Interview with Nisha Charkoudian and Gabrielle Giersch of U.S. Army Research Institute of Environmental Medicine, October 22, 2022.

79 **limits of livability:** Jennifer Vanos et al., "A Physiological Approach for Assessing Human Survivability and Liveability to Heat in a Changing Climate," *Nature Communications* 14 (2023): 7653.

79 **stationary cyclists in Copenhagen:** Bodil Nielsen et al., "Heat Balance During Exercise in the Sun," *European Journal of Applied Physiology and Occupational Physiology* 58 (1988): 189–96.

79 **Florence and Guangzhou:** Marco Morabito et al., "Heat-Related Productivity Loss: Benefits Derived by Working in the Shade or Work-Time Shifting," *International Journal of Productivity and Performance Management* 70, no. 3 (2021): 507–25.

79 **seed packers in Indonesian rainforests:** Yuta Masuda et al., "Warming from Tropical Deforestation Reduces Worker Productivity in Rural Communities," *Nature Communications* 12 (2021): 1601.

80 **men in their thirties and forties:** Julia Belluz, "The Disturbing Hypothesis for the Sudden Uptick in Chronic Kidney Disease," *Vox,* September 4, 2019.

80 **shade in the fields:** Theo Bodin et al., "Intervention to Reduce Heat

Stress and Improve Efficiency Among Sugarcane Workers in El Salvador: Phase 1," *Occupational & Environmental Medicine* 73 (2016): 409–16.

80 **four-post Saran mesh canopies:** Interview with Jason Glaser of La Isla Network, September 8, 2023.

80 **kidney injuries dropped by 70 percent:** Jason Glaser et al., "Preventing Kidney Injury Among Sugarcane Workers: Promising Evidence from Enhanced Workplace Interventions," *Occupational & Environmental Medicine* 77 (2020): 527–34.

80 **Central Valley farmworkers:** Sally Moyce et al., "Heat Strain, Volume Depletion and Kidney Function in California Agricultural Workers," *Occupational & Environmental Medicine* 74 (2017): 402–09.

80 **Amadeo Sumano, a forty-four-year-old:** Interview on August 10, 2021.

81 **doesn't actually happen:** Interview with Juanita Perez, August 24, 2021.

81 **fewer than two hundred safety inspectors:** Former inspector Garrett Brown maintains staffing records at his website Inside Cal/OSHA: https://insidecalosha.org/staffing/.

82 **UFW and the surviving family:** *Bautista v. Cal. Dep't of Occupational Safety & Health,* No. BC-418871 (L.A. Cnty. Super. Ct. Jan. 29, 2009).

82 **"voluntary dehydration":** Edward Adolph, *Physiology of Man in the Desert* (New York: Interscience Publishers, 1947).

83 **died in boot camp:** Grahame M. Budd, "Wet-Bulb Globe Temperature (WBGT)—Its History and Its Limitations," *Journal of Science and Medicine in Sport* 11 (2008): 20–32.

83 **sunlight can raise heat index values:** National Weather Service, "What is the Heat Index?," n.d., accessed November 1, 2024, https://www.weather.gov/ama/heatindex.

84 **what the military calls "acclimatization":** Headquarters, Department of the Army and Air Force, *Heat Stress Control and Heat Casualty Management,* Technical Bulletin, Medical 507, TB MED 507/AFPAM 48-152 (Falls Church, Va.: Office of the Surgeon General, March 2003).

84 **Operation Desert Storm:** U.S. Army Research Institute of Environ-

mental Medicine, *Sustaining Health and Performance in the Desert: A Pocket Guide to Environmental Medicine for Operations in Southwest Asia,* Technical Note 91-2 (Natick, Mass.: U.S. Army Research Institute, December 1990).

84 **heat of battle:** Interview with David Accetta of the U.S. Army Combat Capabilities Development Command Soldier Center, October 22, 2022.

84 **commanders' discipline:** Matthew J. Reardon et al., "Applications of Predictive Environmental Strain Models," *Military Medicine* 162, no. 2 (1997): 136–40.

84 **the world of work:** U.S. Department of Health and Human Services, Public Health Service, Centers for Disease Control, National Institute for Occupational Safety and Health, *Occupational Exposure to Hot Environments (Revised Criteria 1986),* NIOSH Publication Number 86–113 (Washington, D.C.: Government Printing Office, April 1986).

85 **Deepwater Horizon oil spill:** David Michaels and John Howard, "Review of the OSHA-NIOSH Response to the Deepwater Horizon Oil Spill: Protecting the Health and Safety of Cleanup Workers," *PLoS Currents* (July 2012).

85 **ten or twenty minutes:** Ariel Wittenberg and Zack Colman, "Regulators Refuse to Step in as Workers Languish in Extreme Heat," *Politico,* August 8, 2021.

85 **suffered heatstroke:** U.S. House of Representatives, Committee on Education and Labor, Subcommittee on Workforce Protections, *From the Fields to the Factories: Preventing Workplace Injury and Death from Excessive Heat,* 116th Congress, 1st sess., July 11, 2019, 12.

86 **"Many of you will never":** Jim Wasserman, "Summit on Heat: Farmworker Deaths Prompt Regulators to Brave Fiery Sun and Look for Solutions," *The Sacramento Bee,* July 29, 2005.

86 **Los Angeles librarians:** "Worksafe! Indoor Heat Stress Timeline," May 15, 2007.

86 **tomato picker who was crushed:** Robert Salladay and Nancy Vogel, "Gov. Orders Shade, Water for Workers Sickened by Heat," *Los Angeles Times,* August 3, 2005.

86 **Anne Katten, a work safety specialist:** Interview on May 26, 2021.

86 **"In some ways, it's easier":** Interview with Judy Chu, September 7, 2021.

87 **"brought shame to the Valley":** "Death in the Fields," *The Sacramento Bee,* July 29, 2005.

87 **the movie *Predator:*** Salladay and Vogel, "Gov. Orders Shade."

87 **wanted to continue the custom:** Kashkooli interview.

87 **The state senator:** Interview with Dean Florez, August 10, 2021.

89 **"I think it's critical":** California State Senate Democratic Caucus, "Sen. Florez Gathers Officials for 'Meeting in the Sun,'" press release, July 25, 2005.

89 **really an opportunity:** Wasserman, "Summit on Heat."

90 **participate in public rulemaking:** Descriptions of the debates held between 2005 and 2014 regarding the California Division of Occupational Safety and Health's heat illness prevention regulation are drawn from notes and summaries of public hearings and advisory committee meetings convened by the California Occupational Safety and Health Standards Board; public comments submitted to the board; transcripts of California legislature hearings; and contemporaneous articles by Kevin Thompson in *Cal-OSHA Reporter* and Susan Ferriss in *The Sacramento Bee.* These materials were shared with me by Anne Katten of the California Rural Legal Assistance Foundation, Frances Schreiber of Worksafe, and Jordan Segall of Munger, Tolles & Olson.

92 **model of success:** Juanita Constible, "California Lessons for Federal & State Workplace Heat Rules," *Expert Blog,* Natural Resources Defense Council, September 14, 2022, https://www.nrdc.org/bio/juanita-constible/california-lessons-federal-state-workplace-heat-rules.

92 **about two-thirds have plans:** Robert J. Lopez, "California Cuts Back on Safety Enforcement as Farmworkers Toil in Extreme Heat," *Los Angeles Times,* August 14, 2024.

92 **heat injuries have decreased:** Jana Cholakovska and Nate Rosenfield, "Workers Are Dying from Extreme Heat. Why Aren't There Laws to Protect Them?," *Grist,* October 19, 2023.

92 **aren't good enough:** Liza Gross, "For Farmworkers, Heat Too Often Means Needless Death," *Inside Climate News,* July 9, 2021.

92 **At least seventeen workers:** Lopez, "California Cuts Back on Safety Enforcement."

93 **360,000 compensation claims:** Christopher Flavelle, "Work Injuries Tied to Heat Are Vastly Undercounted, Study Finds," *The New York Times,* July 17, 2021.

93 **Stefanie Diaz, a twenty-eight-year-old:** Interview on August 20, 2021.

94 **added Primitivo Cruz:** Interview on August 20, 2021.

94 **water cannot ward off:** Interview with Thomas Bernard of University of South Florida, July 28, 2021.

94 **seriously understaffed:** Brian Edwards and Jacob Margolis, "Higher Temperatures and Less Oversight Mean Workers Are at a Growing Risk in the Climate Emergency," *LAist,* August 25, 2021.

94 **does not require shade for every worker:** Interview with David Hornung of the California Division of Occupational Safety and Health, September 29, 2021.

94 **"never seen it happen":** Perez interview.

95 **piece-rate system that drives workers to death:** Gross, "Heat Too Often Means Needless Death."

95 **Oregon requires more paid rest breaks:** Rachel Spacek, "A Tale of Two Workers: Heat Deaths on the Job Provoke Differing Responses in Oregon, Idaho," *InvestigateWest,* August 15, 2024.

Chapter 5: The Shady Divide

97 **The Shady Divide:** The title is borrowed from a June 15, 2021, *National Geographic* article by Alejandra Borunda: "Los Angeles Confronts Its Shady Divide."

97 **Debbie Stephens-Browder woke up:** Interviews on April 24 and September 26, 2021. Stephens-Browder no longer works for TreePeople.

99 **chase the sun:** Lyra Kilston, *Sun Seekers: The Cure of California* (Los Angeles: Atelier Editions, 2019).

99 **L.A.'s modern identity:** Interview with Mark Vallianatos, former director of LAplus, June 20, 2017.

99 **Freestanding single family homes:** Andrew Whittemore, "How the

Federal Government Zoned America: The Federal Housing Administration and Zoning," *Journal of Urban History* 39, no. 4 (2012): 620–42.

99 **"their shadows fall":** City of Los Angeles, *Department of Public Works Annual Report, 1958–59*, 5.

100 **L.A. planners required:** Interview with Shana Bonstin of the City of Los Angeles Department of City Planning, March 9, 2018. According to the 2006 *L.A. City CEQA Thresholds Guide,* "shadow-sensitive" sites that trigger "mitigation" of nearby buildings include outdoor areas of homes, schools, long-term care facilities, malls, restaurants, and plant nurseries. In 2019, the city removed shading from its default checklist of environmental impacts.

100 **veto new apartment complexes:** Linh Tat, "Loophole Lets Developers Put Big Apartment Buildings Next to SF Valley Houses," *Los Angeles Daily News,* October 2, 2023.

100 **cool and lush landscape:** Vittoria Di Palma and Alexander Robinson, "Willful Waters," *Places,* May 2018.

100 **great forests of oak:** Philip J. Ethington et al., *Historical Ecology of the Los Angeles River Watershed and Environs* (Los Angeles: University of Southern California Spatial Sciences Institute, June 15, 2020).

100 **Spanish cleared these woods:** Ethington et al., *Historical Ecology.*

100 **summer highs dropped:** U.S. Environmental Protection Agency, Office of Policy Analysis, Climate Change Division, *Cooling Our Communities: A Guidebook on Tree Planting and Light-Colored Surfacing,* eds. Hashem Akbari et al. (Washington, D.C.: Government Printing Office, 1992), 13.

101 **Mike Davis, the celebrated scholar:** Interview on March 1, 2018.

101 **about a degree:** U.S. Environmental Protection Agency, *Cooling Our Communities,* 13.

101 **Look at what happened to Pershing Square:** Sam Bloch, "Shade," *Places,* April 2019.

102 **"nuts" and "blabbers":** Timothy G. Turner, "A Tear for Old Pershing Square," *Los Angeles Times,* January 24, 1951.

102 **clear out the "rude panhandlers":** "Pershing Park Adopted by Commission," *Los Angeles Times,* January 17, 1964.

102 **"accosted by derelicts and 'bums'":** "New Design Demanded for Pershing Square," *Los Angeles Times,* January 4, 1964.

102 **"pedestrians will be walking":** "Revised Plan for Pershing Square," *Los Angeles Times,* January 20, 1964.

102 **"see-through" Pershing Square:** Lawrence Culver, *The Frontier of Leisure: Southern California and the Shaping of Modern America* (New York: Oxford University Press, 2010), 79.

102 **"Out went sweet shade":** Art Seidenbaum, "Searching for Square Roots," *Los Angeles Times,* April 17, 1978.

103 **Grand Park offers scant respite:** Interview with architect John Kaliski, March 3, 2018. On the inability to support a shade structure, see Jeff Moravec, "Flying the 'Paper Airplane,'" *Specialty Fabrics Review,* January 1, 2017.

103 **"They like the events":** Interview with Mia Lehrer, February 27, 2018.

103 **new renovation of Pershing Square:** The redesign is now led by Gruen Associates. The Los Angeles–based firm is preserving elements of Agence Ter's proposal including the pergola. See Christian Martinez, "Overhaul of Pershing Square, 'Long a Concrete Jungle' in Downtown L.A., Finally Breaks Ground," *Los Angeles Times,* September 1, 2023; Larry Gordon, "Opinion: L.A.'s Pershing Square Is Getting Another Makeover. Will This Time Finally Be the Charm?," *Los Angeles Times,* April 11, 2024.

103 **landscape designer Lauren Hamer:** Interview with Lauren Hamer and Annelies De Nijs, October 4, 2018. Hamer and De Nijs no longer work for Agence Ter.

104 **green infrastructure of street trees:** Los Angeles recognized its urban forest as city infrastructure in the 1996 General Plan Framework Element.

105 **unchecked development:** Interview with Greg Spotts, formerly of the City of Los Angeles Department of Public Works, Bureau of Street Services, July 16, 2021.

105 **Onerous and outdated regulations:** Laura Messier et al., "Elevating Street Trees to Infrastructure Status: A Comparison of Street Tree Spacing Guidelines in Los Angeles with U.S. Peer Cities," *Urban Forestry & Urban Greening* 103 (January 2025).

105 **one-quarter of 1 percent:** Los Angeles allotted an estimated $25.4 million on urban forestry in fiscal year 2018–19; at the time, the city's adopted budget was $9.9 billion. See Dudek, *First Step: Developing an Urban Forest Management Plan for the City of Los Angeles* (Los Angeles: City Plants, 2018). L.A.'s paltry funding does not surprise Stephanie Pincetl, a UCLA environmental studies professor. "We think about trees as nature, and nature is perceived to be free, not something the public invests in to make a system work," she told me in a July 21, 2021, interview. For more durable public shadescapes, Pincetl suggested L.A. could incentivize real estate developers to build arcaded sidewalks, like the porticoes of Bologna.

105 **hired the Olmsted Brothers:** Greg Hise and William Deverell, *Eden by Design: The 1930 Olmsted-Bartholomew Plan for the Los Angeles Region* (Berkeley: University of California Press, 2000).

105 **chartered with a debt limit:** Interview with Marques Vestal, February 27, 2023.

105 **federal funding for public infrastructure:** Mike Davis, *Ecology of Fear: Los Angeles and the Imagination of Disaster* (New York: Henry Holt, 1998), 67–69.

106 **Watts, a hard and gray neighborhood:** Trees cover 14 percent of Watts, compared to 24 percent of the city overall. See TreePeople, Los Angeles County Tree Canopy Map Viewer, n.d., accessed November 20, 2024, https://treepeople.org/los-angeles-county-tree-canopy-map-viewer/.

106 **"it is ninety-five here":** Watts has temperatures about five degrees above the Los Angeles average. See UCLA Luskin Center for Innovation, *Watts Rising: 2024 Progress Report on Implementation of the Transformative Climate Communities Program Grant* (Regents of the University of California, Los Angeles, July 2024), 26.

106 **trees and other environmental amenities:** Jeremy S. Hoffman et al., "The Effects of Historical Housing Policies on Resident Exposure to Intra-Urban Heat: A Study of 108 US Urban Areas," *Climate* 8, no. 1 (January 13, 2020).

106 **formerly redlined neighborhoods:** David J. Nowak et al., "The Disparity in Tree Cover and Ecosystem Service Values Among Redlining

Classes in the United States," *Landscape and Urban Planning* 221 (May 2022).

107 **road widening to ease traffic:** Hise and Deverell, *Eden by Design,* 114; Michael Manville, "Automatic Street Widening: Evidence from a Highway Dedication Law," *Journal of Transport and Land Use* 10, no. 1 (2017): 375–93.

107 **safer with trees:** Jody Rosenblatt Naderi et al., "The Street Tree Effect and Driver Safety," *ITE Journal on the Web* (February 2008): 69–73.

107 **overhead power lines:** Han Fu and Qiaoqi Dai, "Short on Shade: Research on Equity and Exposure in Los Angeles," SWA Group, 2023.

107 **fewer than a dozen:** Interview with Alejandro Fabian of TreePeople, April 26, 2021.

107 **twenty-one years to trim a tree:** Kenneth Mejia, Los Angeles city controller, *Follow-Up: L.A. Tree-Trimming & Maintenance Audit,* April 17, 2023.

107 **four days to fill a pothole:** Carla Hall, "Commentary: So Many Potholes in L.A.—and Not Enough People to Fix Them," *Los Angeles Times,* April 9, 2024.

107 **Scott Goldstein, a Hollywood screenwriter:** Interview on September 12, 2021.

108 **not typically funded for maintenance:** Blanca Begert, "Cities Want More Trees. Drought Is Complicating Their Efforts," *Grist,* October 18, 2022. This is not unique to Los Angeles. As a Baltimore urban forester lamented, "There is no funding for maintenance because you are handing over a shit ton of money without any deliverables." Mariya Shcheglovitova, "Valuing Plants in Devalued Spaces: Caring for Baltimore's Street Trees," *Environment and Planning E: Nature and Space* 3, no. 1 (March 2020): 228–45.

108 **imperative in L.A.:** Interview with David Nowak of the U.S. Forest Service, May 18, 2022.

108 **most residents pass:** Historically, only 10 percent of contacted residents agree to adopt street trees. Katherine A. McNamara et al., "A Novel Resident Outreach Program Improves Street Tree Planting Outcomes in Los Angeles," *Environmental Challenges* 9 (December 2022): 100596.

108 **destroying the pavement:** "Editorial: L.A. Tripped over Its Duty to Fix the City's Broken Sidewalks," *Los Angeles Times,* November 23, 2021. Sidewalk damage can be mitigated by root pruning, a regular practice in nearby Santa Monica, according to a September 16, 2021, interview with city foresters Matthew Wells and Wister Dorta. See also Greg McPherson, "Expenditures Associated with Conflicts Between Street Tree Root Growth and Hardscape in California, United States," *Journal of Arboriculture* 26, no. 6 (November 2000): 289–97.

108 **adjacent property owner:** Los Angeles Bureau of Street Services, Urban Forestry FAQs, n.d., accessed November 20, 2024, https://streetsla.lacity.org/faqs-ufd.

108 **it's a lousy policy:** Stephanie Pincetl, "From the Sanitary City to the Sustainable City: Challenges to Institutionalising Biogenic (Nature's Services) Infrastructure," *Local Environment: The International Journal of Justice and Sustainability* 15, no. 1 (2010): 43–58. Researchers in Detroit found that residents rejected new street trees because the city failed to maintain the ones that were already there. See Christine E. Carmichael and Maureen H. McDonough, "Community Stories: Explaining Resistance to Street Tree-Planting Programs in Detroit, Michigan, USA," *Society & Natural Resources* 32, no. 5 (January 2019): 588–605; Lester Graham, "Feds Spend Big on Urban Tree-Planting in Michigan. Will Residents Get a Say?," *Bridge Michigan,* July 2, 2024.

109 **During a heat wave, Watts residents:** UCLA Center for Healthy Climate Solutions and UCLA Center for Public Health and Disasters, UCLA Heat Maps, n.d., accessed November 20, 2024, https://sites.google.com/g.ucla.edu/uclaheatmaps/map.

109 **elderly Black and Latino residents:** Los Angeles Urban Cooling Collaborative, *Rx for Hot Cities 2: Reducing Heat and ER Visits with Trees and High-Albedo Surfaces in Los Angeles* (Beverly Hills: TreePeople, 2023).

109 **"climate gap":** Seth B. Shonkoff et al., "The Climate Gap: Environmental Health and Equity Implications of Climate Change and Mitigation Policies in California—a Review of the Literature," *Climatic Change* 109 (2011): 485–503.

109 **annual provision of $25 million:** Dudek, *First Step.*

110 **at least $664 million:** CAPA Strategies, *Los Angeles Urban Forest Equity Streets Guidebook,* April 2021.

110 **$175 million spent every year:** "Pavement preservation" and sewer construction and maintenance funding figures taken from the City of Los Angeles 2023–24 Budget Summary booklet.

110 **"Most people don't like":** Interview with Eric Garcetti, October 4, 2021.

110 **reduce L.A.'s infamous car dependency:** Ian Lovett, "A Los Angeles Plan to Reshape the Streetscape Sets Off Fears of Gridlock," *The New York Times,* September 7, 2015.

111 **"Maybe you haven't thought":** Tim Arango, "'Turn Off the Sunshine': Why Shade Is a Mark of Privilege in Los Angeles," *The New York Times,* December 1, 2019.

111 **shade 750 forlorn bus stops:** Correspondence with Christopher Hawthorne, October 2019.

111 **ninety thousand new trees:** Eric Garcetti, mayor of Los Angeles, *L.A.'s Green New Deal: Sustainable City Plan,* 2019.

111 **would not achieve the mayor's definition:** CAPA Strategies, *Equity Streets Guidebook.*

111 **L.A.'s transportation plan:** Kerry Cavanaugh, "Commentary: L.A.'s Promise for Safer Streets Has Stalled. But a Ballot Measure Could Restart the Mobility Plan," *Los Angeles Times,* February 15, 2022.

112 **Felipe Escobar, a community organizer:** Interview on July 21, 2021. Escobar has since left Pacoima Beautiful.

112 **prolongs the exposure:** According to the Centers for Disease Control, Americans who do not own cars "may be more likely to be exposed to extreme heat in the course of daily activities." See Kea Wilson, "This Heat Wave Is a Car Dependency Problem," *Streetsblog USA,* July 18, 2024.

112 **seven times likelier:** Todd Woody, "How One Los Angeles Neighborhood Is Guarding Against Deadly Heat," *Bloomberg,* July 25, 2022.

113 **elderly woman named Olga Hernandez:** Interview on October 4, 2021.

114 **I first heard it from Tim Watkins:** Interview on June 22, 2017.

114 **"Watts needed to have":** Warren Olney, "In Our Backyard No. 3: Heat Is the Deadliest Aspect of Climate Change. It's Turning Some Underserved LA Neighborhoods Red Hot," *To The Point* (podcast), May 6, 2021.

115 **Requests to deforest:** LAPD senior lead officers received or submitted at least 130 requests to remove or trim trees for public safety in 2020, according to emails I obtained through public records requests to the police department.

115 **"It's not that they have the authority":** Interview with Michael Pinto on June 5, 2017.

115 **Susan Phillips, a Pitzer College anthropologist:** Interview on September 2, 2021.

116 **Since 1995:** Los Angeles Police Department, Design Out Crime, n.d., accessed November 20, 2024, https://www.lapdonline.org/design-out-crime/.

116 **guide their safety assessments:** In 2020, the police department's Community Safety Partnership Bureau briefed housing managers on CPTED principles and urged increased visibility in seven sites, according to public records I received. The department also similarly assessed two public parks.

116 **Perry Crouch, a longtime Watts resident:** Interview on September 16, 2021.

116 **Universal Studios hat-racked:** Jeanette Marantos, "Severe Pruning During Summer Heat Is Hell for Trees, Not Just Striking Actors," *Los Angeles Times,* July 21, 2023.

116 **The core principles of CPTED:** Oscar Newman, *Defensible Space: Crime Prevention Through Urban Design* (New York: Macmillan, 1972).

117 **"access control":** The International CPTED Association (ICA), "Primer in CPTED—What is CPTED?," n.d., accessed December 10, 2024, https://www.cpted.net/primer-in-cpted.

117 **"negative message about trees":** Richard Conniff, "Trees Shed Bad Rap as Accessories to Crime," Yale School of Forestry and Environmental Studies, 2012.

117 **prefer not to have them:** Interview with Lisa Sarno, formerly of Million Trees L.A., August 16, 2021.

117 **"Natural surveillance is easy":** Interview with James Nichols and Alfonso Velasco, September 30, 2021.

117 **Heat is an irritant:** Craig A. Anderson, "Heat and Violence," *Current Directions in Psychological Science* 10, no. 1 (2001): 33–38.

118 **A minor altercation:** Ajai Raj, "Feeling Hot Can Fuel Rage: Hotter Weather Sparks Aggression and Revolution," *Scientific American,* January 1, 2014.

118 **Reams of associational studies:** Clayton Page Aldern, "Climate Change Won't Just Make the World a Hotter Place—It Will Make It a More Violent One, Too," *The Globe and Mail,* April 5, 2024.

118 **trial "novel solutions":** Kilian Heilmann et al., "Op-Ed: How Cities Can Break the Link Between Heat and Crime and Save Lives," *Los Angeles Times,* July 29, 2021.

118 **Kelly Turner, a UCLA urban planning:** Steve Chiotakis, "Using—and Removing—Shade as a Political Tool," *Greater LA* (podcast), July 26, 2023.

118 **Motorola video cameras:** Richard Winton, "LAPD to Train Cameras on High-Crime Area in Watts," *Los Angeles Times,* September 23, 2005.

118 **recommending tree removals:** Letter from William A. Robertson of the Bureau of Street Services to Bill Rosendahl of the Public Works Committee, February 14, 2007, Los Angeles City Council File 06-2445.

118 **South L.A.'s Harvard Park:** Interview with Wayne Caffey, formerly of the LAPD, June 21, 2017.

119 **Aaron Thomas, a veteran urban forester:** Karl Krause, "Whose Eyes on the Street?," *Landscape Architecture Magazine,* May 11, 2021.

119 **"law enforcement associates safety":** Interview with Cynthia Gonzalez, September 27, 2021.

119 **New York City's urban forest:** New York City Housing Authority, Capital Projects Division, *NYCHA's Urban Forest: A Vital Resource for New York City,* October 2021.

119 **In Chicago, a much-cited 2001 study:** Frances E. Kuo and Wil-

liam C. Sullivan, "Environment and Crime in the Inner City: Does Vegetation Reduce Crime?," *Environment and Behavior* 33, no. 3 (2001): 343–67.

120 **university researchers in Baltimore and Portland:** Conniff, "Trees Shed Bad Rap."

121 **fourteen thousand fires:** Joel Grover, "Up in Flames: Rising Number of Homeless Fires Threatens LA Neighborhoods," NBC4 Los Angeles, May 9, 2024.

121 **"Nothing I can do legally":** Emails from L.A. police officers Jose Moreno, Jonathan Ojeda, and Sandra Zamora, and L.A. City Council staffers Tim Glick, Colin Crews, and Michael Owens pertaining to tree trimming and removals were obtained through public records requests.

122 **Elizabeth Chou, a reporter:** "An LAPD Officer in the Valley Wanted to Displace Unhoused People by Removing a Tree's Shade," *Los Angeles Public Press,* July 18, 2023. Chou additionally reported from a community meeting at de Toledo High School on Twitter (@reporterliz) on July 11, 2023.

122 **police watchdogs decry:** Stop LAPD Spying Coalition, *Automating Banishment: The Surveillance and Policing of Looted Land,* November 2021.

122 **The sun dehydrates:** Interview with Michele Seckington of the VA Greater Los Angeles Healthcare System, June 28, 2021.

123 **Coley King, a doctor:** Interview on June 20, 2021.

123 **In Los Angeles, the unhoused account:** Summer Lin, " 'He Baked': Heat Waves Are Killing More L.A. Homeless People Who Can't Escape Broiling Sun," *Los Angeles Times,* February 19, 2023.

123 **two to three hundred times:** Katherine Davis-Young, "Homelessness Is Aggravating Harm Caused by the Phoenix Heat, Medical Personnel Say," NPR, August 30, 2022.

123 **Debra McNelley, one of the residents:** Chou tweeted an audio interview with McNelley and a written statement in separate posts on July 11, 2023.

125 **Watts Rising, an effort to upgrade:** UCLA Luskin Center for Innovation, *Watts Rising.*

125 **Helene Schpak:** Bloch, "Shade."

Chapter 6: The Heat Dome

127 **It came from the other side:** James Ross Gardner, "Seventy-Two Hours Under the Heat Dome," *The New Yorker,* October 11, 2021.

127 **Pacific Northwest's driest since 1924:** National Integrated Drought Information System (NIDIS), "Drought Status Update for the Pacific Northwest," July 29, 2021.

128 **National Weather Service hesitated:** Aaron Mesh, "A 'Heat Dome' Could Send Portland Temperatures to 110 Degrees Next Week," *Willamette Week,* June 21, 2021.

128 **On Wednesday, June 23:** Multnomah County, "From the Archives: Forecast Calls for 'Life-Threatening' Heat; Health Officer Urges Public to Make a Plan, Help Others," June 23, 2021, https://multco.us/news/archives-forecast-calls-life-threatening-heat-health-officer-urges-public-make-plan-help.

128 **on Friday, June 25:** The descriptions of heat casualties over the weekend are based on transcripts of 911 calls and computer-aided dispatch (CAD) reports published by the Portland Bureau of Emergency Communications (BOEC) between June 25 and June 30, 2021, obtained through a public records request, as well as Gardner, "Under the Heat Dome."

129 **The Rose City was shutting down:** Aron Yohannes, "Portland Heat Wave Leads to Closures, Cancellations and Schedule Changes," *The Oregonian/OregonLive,* June 25, 2021; Ardeshir Tabrizian, "MAX, WES Rail Service Suspended; So Is Ban on Pumping Your Own Gas," *The Oregonian/OregonLive,* June 28, 2021; Sophie Peel, "Residents in Apartments Without Air Conditioning Describe Fear and Frustration During Heat Wave," *Willamette Week,* July 3, 2021.

129 **fifty-foot gash:** Sophie Peel, "As Portland Heat Hits 116 Degrees, Road Buckles in Kenton Neighborhood," *Willamette Week,* June 29, 2021.

129 **On Sunday, June 27:** Ardeshir Tabrizian, "Over 6,000 PGE Customers Lost Power Amid Record Heat Wave," *The Oregonian/OregonLive,* June 27, 2021; Ardeshir Tabrizian, "PGE Customers Twice Broke Electricity Usage Records During June Heat Wave," *The Oregonian/OregonLive,* July 29, 2021; Dave Jamieson, "Oregon Workers Feared

for Their Lives During Heat Wave, OSHA Complaints Show," *Huff-Post,* July 16, 2021; Amelia Templeton and Sam Stites, "Oregon Wasn't Prepared for the Heat Wave. Experts Say the State Can Do Better," Oregon Public Broadcasting, July 10, 2021.

129 **"There's a lore":** Interview with Vivek Shandas, August 5, 2021.

129 **On Monday, June 28:** Multnomah County Communications Office, "News Release: Multnomah County Medical Examiner Finds 45 Deaths Related to Historic Heat Wave," June 30, 2021; Cristin Severance, "36-Year-Old Woman Died During Oregon Heat Wave After Getting Dropped Off by Medical Transport," KGW News, August 12, 2021; Maxine Bernstein, "Lack of Air Conditioning, Wife's Infirmities Led to Her Heat-Related Death in SE Portland Home, Husband Says," *The Oregonian/OregonLive,* August 8, 2021.

130 **"virtually impossible":** Sjoukje Y. Philip et al., "Rapid Attribution Analysis of the Extraordinary Heat Wave on the Pacific Coast of the US and Canada in June 2021," *Earth System Dynamics* 13 (2022): 1689–713.

130 **Omega blocks:** Michael E. Mann and Susan Joy Hassol, "That Heat Dome? Yeah, It's Climate Change," *The New York Times,* June 29, 2021.

130 **it could occur every decade:** Tim Dickinson, "Deadly Pacific Northwest Heatwave: 'Virtually Impossible' Without Man-Made Warming," *Rolling Stone,* July 8, 2021.

130 **In a typical year in metro Portland:** Multnomah County, *Final Report: Health Impacts from Excessive Heat Events in Multnomah County, Oregon, 2021* (June 2022).

130 **than from Covid-19:** "Tracking Coronavirus in Multnomah County, Ore.: Latest Map and Case Count," *The New York Times,* n.d., https://www.nytimes.com/interactive/2021/us/multnomah-oregon-covid-cases.html.

130 **1,200 lives in the United States and Canada:** John Ryan, "Can Seattle Take the Heat? Officials Say Area Is Better Prepared This Summer," KUOW, June 22, 2023.

131 **drive around the city:** Shandas's journey was chronicled by Sophie Peel, "This Is the Hottest Place in Portland," *Willamette Week,* July 14, 2021; Gardner, "Under the Heat Dome"; Nadina Galle, *The Nature of*

Our Cities: Harnessing the Power of the Natural World to Survive a Changing Planet (New York: Mariner Books, 2024), 93–118.

131 **Heatproof homes:** Alexandra R. Rempel et al., "Improving the Passive Survivability of Residential Buildings During Extreme Heat Events in the Pacific Northwest," *Applied Energy* 321 (September 2022): 119323.

131 **the city had dragged boulders:** Samantha Swindler et al., "Where Are Portland's Urban Heat Islands?," posted July 29, 2021, by *The Oregonian*, YouTube, https://www.youtube.com/watch?v=u3KWIrEP9AI; Galle, *Nature of Our Cities*, 98.

133 **In the aftermath:** Multnomah County, *Final Report.*

133 **studied two hundred years ago by Luke Howard:** Gerald Mills, "Luke Howard and *The Climate of London*," *Weather* 63, no. 6 (June 2008): 153–57.

134 **"The heat island is mitigated":** Helmut E. Landsberg, "Atmospheric Changes in a Growing Community (The Columbia, Maryland Experience)," *Urban Ecology* 4, no. 1 (May 1979): 53–81.

135 **Mayor John Lindsay established:** Michael Hebbert and Vladimir Jankovic, "Cities and Climate Change: The Precedents and Why They Matter," *Urban Studies* 50, no. 7 (May 2013): 1332–47.

135 **National Institutes of Health conference:** Douglas H. K. Lee, editorial, *Environmental Research* 5, no. 1 (March 1972): i–ii; Frank P. Ellis, "Mortality from Heat Illness and Heat-Aggravated Illness in the United States," *Environmental Research* 5, no. 1 (March 1972): 1–58; Stanley H. Schuman, "Patterns of Urban Heat-Wave Deaths and Implications for Prevention: Data from New York and St. Louis During July, 1966," *Environmental Research* 5, no. 1 (March 1972): 59–75.

136 **urban heat implications of redlining:** Hoffman et al., "Effects of Historical Housing Policies"; Nowak et al., "Disparity in Tree Cover."

136 **a mere correlation:** Dave Miller, "PSU Professor Suggests Ways Cities Could Adapt to Hotter Temperatures," *Think Out Loud* (podcast), August 13, 2021.

137 **Shandas has pursued:** Vivek Shandas, "Climate Justice: Towards a Proactive Response to Social Inequities," *Connections* 10, no. 11 (2009): 4–5.

138 **tearing down vacant buildings:** Nowak interview.

138 **Why not heat overlays:** Miller, "PSU Professor."

138 **collect fifty thousand street-level temperature readings:** Jackson Voelkel and Vivek Shandas, "Towards Systematic Prediction of Urban Heat Islands: Grounding Measurements, Assessing Modeling Techniques," *Climate* 5, no. 2 (2017).

138 **"when you actually go down":** Jim Morrison, "Can We Turn Down the Temperature on Urban Heat Islands?," *Yale Environment 360,* September 12, 2019.

139 **spare more room on the ground:** Interview with Vivek Shandas, August 21, 2024. At least since the 1960s, West Coast environmentalists have suggested vertical development as a preservation tactic. See Davis, *Ecology of Fear,* 83.

139 **urban corridor that has been rewilded:** Yasuyo Makido et al., "Nature-Based Designs to Mitigate Urban Heat: The Efficacy of Green Infrastructure Treatments in Portland, Oregon," *Atmosphere* 10, no. 5 (2019).

139 **"None of it is mandatory":** Shandas interview, August 2024.

140 **fallen on deaf ears:** Michelle Nijhuis, "When Summer Becomes the Season of Danger and Dread," *The New Yorker,* September 10, 2022.

140 **Multnomah County has acknowledged:** Peel, "Hottest Place."

140 **They raced to do what they could:** Rob David et al., "Oregon's Deadly Heat: 'There Will Be a Next Time,'" *The Oregonian/OregonLive,* July 4, 2021; Sarah Kaplan, "Heat Waves Are Dangerous. Isolation and Inequality Make Them Deadly," *The Washington Post,* July 21, 2021; Jessica Guernsey and Jennifer Vines, *Preliminary Review on Excessive Heat Deaths, Multnomah County, June 2021* (July 2021).

141 **deaths were more numerous:** Peel, "Hottest Place."

141 **The fatality rate in Lents:** Jack Forrest, "ZIP Code for Lents Neighborhood Had Highest Concentration of Multnomah County Heat Victims," *The Oregonian/OregonLive,* July 3, 2021.

141 **"I've been saying":** Shandas interview, August 2021.

142 **eleven million Americans:** Andre M. Perry and David Harshbarger, "America's Formerly Redlined Neighborhoods Have Changed, and So Must Solutions to Rectify Them," Brookings Institute, October 14, 2019.

142 **10-degree increase in temperature:** G. Brooke Anderson et al., "Heat-Related Emergency Hospitalizations for Respiratory Diseases in the Medicare Population," *American Journal of Respiratory and Critical Care Medicine* 187, no. 10 (May 2013): 1098–103.

142 **In Richmond, Virginia:** Peter Braun et al., "A Heat Emergency: Urban Heat Exposure and Access to Refuge in Richmond, VA," *GeoHealth* 8, no. 6 (June 2024): e2023GH000985.

142 **residents stranded on redlined heat islands:** Ryan Best and Elena Mejia, "The Lasting Legacy of Redlining," FiveThirtyEight, February 9, 2022.

142 **A study of Sacramento neighborhoods:** Jared M. Ulmer et al., "Multiple Health Benefits of Urban Tree Canopy: The Mounting Evidence for a Green Prescription," *Health & Place* 42 (November 2016): 54–62.

142 **fewer heart attacks:** Scott C. Brown et al., "Longitudinal Impacts of High Versus Low Greenness on Cardiovascular Disease Conditions," *Journal of the American Heart Association* 13, no. 19 (October 2024): e029939.

142 **usually located in white neighborhoods:** Robert I. McDonald et al., "The Tree Cover and Temperature Disparity in US Urbanized Areas: Quantifying the Association with Income Across 5,723 Communities," *PLoS One* 16, no. 4 (April 2021): e0249715.

142 **high rates of asthma:** Kaitlyn Adams and Colette Steward Knuth, "The Effect of Urban Heat Islands on Pediatric Asthma Exacerbation: How Race Plays a Role," *Urban Climate* 53 (January 2024): 101833.

142 **44 percent fewer green acres:** Ronda Chapman et al., *Parks and an Equitable Recovery* (San Francisco: The Trust for Public Land, 2021).

143 **easier to maintain than a grassy field:** Roberto J. Manzano, "Valley Roundup: Sherman Oaks: Chandler Elementary Puts Down New Roots," *Los Angeles Times,* August 29, 1999.

143 **more vulnerable to heat:** Eric Kennedy et al., *Thermally Comfortable Playgrounds: A Review of Literature and Survey of Experts* (National Program for Playground Safety at the University of Northern Iowa, 2020), 16.

143 **rate of warming could be double:** Stone, *The City,* 87.

143 **too hot to play:** Shawnda A. Morrison, "Moving in a Hotter World: Maintaining Adequate Childhood Fitness as a Climate Change Countermeasure," *Temperature* 10, no. 2 (2023): 179–97.

143 **Phoenix, Miami, and Los Angeles:** Anne deBoer et al., *Economic Assessment of Heat in the Phoenix Metro Area* (Arlington, Va.: The Nature Conservancy, 2021); Atlantic Council Climate Resilience Center, "Hot Cities, Chilled Economies: Impacts of Extreme Heat on Global Cities," n.d., accessed December 1, 2024, https://onebillionresilient.org/hot-cities-chilled-economies.

144 **"thermally comfortable route":** Maggie Messerschmidt et al., *Heat Action Planning Guide for Neighborhoods of Greater Phoenix* (Arlington, Va.: The Nature Conservancy, 2019).

144 **urban parks in Vancouver and Sacramento:** Rachel A. Spronken-Smith and Timothy R. Oke, "The Thermal Regime of Urban Parks in Two Cities with Different Summer Climates," *International Journal of Remote Sensing* 19, no. 11 (July 1998): 2085–104.

144 **shade out the health risks:** Robert D. Brown et al., "Designing Urban Parks That Ameliorate the Effects of Climate Change," *Landscape and Urban Planning* 138 (June 2015): 118–31.

144 **It takes only three seconds:** Kennedy et al., *Thermally Comfortable Playgrounds,* 18.

144 **drop summer temperatures:** Stone, *Radical Adaptation,* 35–36.

145 **could be 3 degrees lower:** Los Angeles Urban Cooling Collaborative, *Rx for Hot Cities: Climate Resilience Through Urban Greening and Cooling in Los Angeles* (Beverly Hills: TreePeople, 2020).

Chapter 7: Shelter from the Sun

149 **As Portland's Lents neighborhood burned:** Interview with Jeff Stern, July 1, 2021.

150 **passive houses can be safe havens:** According to federal building scientists, homes built to passive house standards in Houston, Atlanta, Los Angeles, Portland, Detroit, and Minneapolis could remain "habitable" for the duration of a seven-day heat wave without power. See Ellen Franconi et al., *Enhancing Resilience in Buildings Through Energy Efficiency* (Richland, Wash.: Pacific Northwest National Laboratory,

July 2023). In real life, a century-old Craftsman bungalow retrofit to passive house standards remained habitable during a 2021 winter grid failure in Austin, Texas. See Stacey Freed, "Age of Resilience: Cold Case," *Green Builder,* September 9, 2021.

150 **German physicist named Wolfgang Feist:** Katrin Krämer, "The First Passive House: Interview with Dr. Wolfgang Feist," *iPHA Blog,* October 19, 2016; Justin Bere, *An Introduction to Passive House* (London: RIBA Publishing, 2019).

151 **roughly a quarter of the rate:** Correspondence with Ken Levenson of the Passive House Network, November 2024.

151 **living room soars to seventeen feet:** Randy Gragg, "An Eco-Chic Passive House," *Portland Monthly* (January 2014).

152 **Building engineers call windows the holes:** Joseph Lstiburek, "BSI-104: Punched Openings," *Insight* 104 (Westford, Mass.: Building Science Corporation, April 2018).

152 **glass allows more heat:** Interview with Steven Bruning of Newcomb & Boyd, November 2, 2021.

152 **Manchester, Edinburgh, and Glasgow:** Susan Roaf, "Are You Sufficiently Prepared for a Heatwave Summer?," *Building Design,* April 17, 2023.

152 **San Francisco, where notoriously cold summers:** Bob Egelko, "S.F. Condo Owners Can Sue Architects," *SFGATE,* December 12, 2012; Emily Landes, "Nearly $10M Settlement for 'Cooked' SF Condo Owners," *The Real Deal,* July 28, 2021.

152 **half of summer cooling loads:** Correspondence with Stephen Selkowitz, November 2021.

152 **building was code compliant:** Landes, "Nearly $10M Settlement."

152 **Architects, engineers, and developers use past conditions:** Dena Prastos discussed architects' "backwards-looking" standards on the *America Adapts* podcast on June 19, 2023: "The Fundamentals (and Ethics) of Architecture and Climate Adaptation."

153 **41 percent of homes were air-conditioned:** Kyle Iboshi, "Number of Portland Homes with Air Conditioning Has Nearly Doubled in Past Decade," KGW.com, July 27, 2022.

153 **Since the days of the Olgyays:** Victor Olgyay, *Design with Climate:*

Bioclimatic Approach to Architectural Regionalism, new and expanded ed. (Princeton, N.J.: Princeton University Press, 2015), 63–74.

155 **Wolfgang Feist commissioned the first *Passivhaus:*** "The World's First Passive House, Darmstadt-Kranichstein, Germany," *Passipedia,* n.d., accessed December 12, 2024, https://passipedia.org/examples/residential_buildings/multi-family_buildings/central_europe/the_world_s_first_passive_house_darmstadt-kranichstein_germany.

155 **Feist's windows are recessed:** Wolfgang Feist, "The Passive House in Darmstadt-Kranichstein During Spring, Summer, Autumn and Winter," Passive House Institute, 2006, https://passiv.de/former_conferences/Kran/Passive_House_Spring_Winter.htm.

156 **balcony is not effective:** The heat gained by south-facing windows on clear days in Feist's latitude was calculated with the Overhang Analysis tool on the Sustainable By Design website: https://www.susdesign.com/overhang.

156 **When the slats are open:** Solar heat gain coefficients for louvered sunscreens and other exterior shading devices are summarized in *Residential Exterior Window Shading Deemed Measure Assessment Study,* an October 2022 study by Opinion Dynamics sponsored by the California Public Utilities Commission to research the market and energy savings potential of shading devices.

156 **London office's cooling load:** Federico Seguro and Jason Palmer, *Solar Shading Impact: Business Case, Strategic Vision, Action Plan for British Blind & Shutter Association* (Milton Keynes, UK: National Energy Foundation, 2016).

156 **reduce AC use by 78 to 94 percent:** Eleanor S. Lee et al., *High Performance Building Façade Solutions: PIER Final Project Report* (Berkeley, Calif.: Lawrence Berkeley National Laboratory, 2009), 11.

156 **Feist installed arrays:** Wolfgang Feist, "30 Years: The World's First Passive House," International Passive House Institute, June 25, 2021, YouTube, https://www.youtube.com/watch?v=gG8WK6CD120.

156 **Shades prove the exception:** In New York, architectural shades can be found on passive houses at 425 Grand Concourse in the Bronx, 803 Knickerbocker Avenue in Brooklyn, and 211 West Twenty-Ninth Street and 470 Columbus Avenue in Manhattan, among others.

157 **use less glass to begin with:** The standard window-to-wall ratio for buildings certified by the Passive House Institute U.S. (2024) is 18 percent, compared to 30 percent in the International Energy Conservation Code (2021) and 40 percent in ASHRAE 90.1 (2022), the two most common energy codes in the United States.

157 **like slow guillotines:** Credit to Edwin Heathcote for this evocative description of motorized venetian blinds ("The Humble Awning Is Ready for Its Time in the Sun," *Financial Times,* June 19, 2023).

157 **less than 1 percent of new American construction:** Patrick Sisson, "Is This the Most Energy-Efficient Way to Build Homes?," *MIT Technology Review,* December 22, 2023.

158 **largest developers resist efforts:** Christopher Flavelle, "Secret Deal Helped Housing Industry Stop Tougher Rules on Climate Change," *The New York Times,* October 26, 2019.

159 **incentives have boosted passive house construction:** Hannah Seo, "The Passive House Trend Is Booming," *The Washington Post,* October 30, 2024.

159 **cost 3 to 5 percent more:** Mary Salmonsen, "Report: Passive House Nears Cost Parity with Traditional Construction," *Multifamily Dive,* August 10, 2023; Teresa Xie, "What 'The Curse' Gets Wrong About Passive House Architecture," *Bloomberg,* February 13, 2024.

159 **sixteen thousand passive house apartments:** Salmonsen, "Passive House Nears Cost Parity."

159 **commanded to reach zero emissions:** Mark Segal, "EU Adopts Rules Requiring All New Buildings to Be Zero Emissions by 2030," *ESG Today,* April 12, 2024.

159 **all but essential in many countries:** Correspondence with Ann van Eycken of the European Solar Shading Organisation, August and September 2023.

159 **"I move into a building":** These comments were made by Alexandria Ocasio-Cortez at a Bronx town hall on May 30, 2019, and live streamed on Facebook by Housing Justice for All (@housing4allNY).

160 **"I'm not sold on the passive house":** Interview with Daniel Barber, December 14, 2021.

160 **passive houses are still high-carbon architecture:** Michael Maines,

"Is There Environmentally Friendly Spray Foam Insulation?," *Fine Homebuilding* 297 (February/March 2021); Fred A. Bernstein, "Taking a Holistic Approach to Embodied Carbon," *Architectural Record,* October 10, 2022.

160 **Buildings account for around 40 percent:** Anna Dyson et al., *Building Materials and the Climate: Constructing a New Future* (Nairobi: United Nations Environment Programme, 2023).

161 **foliage that is so scarce:** Milan's tree canopy cover is 16 percent, according to a 2020 report by Forestami, a greening initiative of the Polytechnic University of Milan and local government agencies.

161 **"People would rather live":** Interview with Anastasia Kucherova, June 24, 2021.

161 **brought swallows to Milan:** Piero Piccardi, "'We Have Brought Swallows into Milan,' Says Father of the Vertical Forest," *Climate Home News,* October 4, 2018.

162 **reduce solar heating at least 40 percent:** Elena Giacomello and Massimo Valagussa, *Vertical Greenery: Evaluating the High-Rise Vegetation of the Bosco Verticale, Milan* (Chicago: Council on Tall Buildings and Urban Habitat, 2015).

162 **Concrete is an environmental nightmare:** Kristoffer Tigue, "Concrete Is Worse for the Climate Than Flying. Why Aren't More People Talking About It?," *Inside Climate News,* June 24, 2022.

162 **The Bosco Verticale requires more:** Lloyd Alter, "Another Look at Stefano Boeri's Vertical Forest," *Treehugger,* n.d., updated August 13, 2020.

163 **never offset the carbon emissions:** Victoria Kate Burrows, "It's Not That Easy Being Green," World Green Building Council (blog), n.d., accessed September 1, 2024, https://worldgbc.org/article/its-not-that-easy-being-green/.

163 **architects in Niger and Burkina Faso:** Peter Schwartzstein, "The Extraordinary Benefits of a House Made of Mud," *National Geographic,* January 19, 2023.

163 **A freshly painted white roof:** George A. Ban-Weiss et al., "Using Remote Sensing to Quantify Albedo of Roofs in Seven California Cities, Part 1: Methods," *Solar Energy* 115 (May 2015): 777–90.

164 **"passive daytime radiative cooling":** Aaswath P. Raman et al., "Passive Radiative Cooling Below Ambient Air Temperature Under Direct Sunlight," *Nature* 515, no. 7528 (November 2014): 540–44.

164 **overwhelmed by the growth in emissions:** U.S. Energy Information Administration, "EIA Projections Indicate Global Energy Consumption Increases Through 2050, Outpacing Efficiency Gains and Driving Continued Emissions Growth," press release, October 11, 2023.

164 **a life "after comfort":** Daniel Barber, "After Comfort," *Log* 47 (Fall 2019): 45–50.

164 **ideal indoor air temperature range:** A. H. Taki et al., "Assessing Thermal Comfort in Ghadames, Libya: Application of the Adaptive Model," *Building Services Engineering Research and Technology* 20, no. 4 (1999): 205–10.

164 **Burkinabe prefer warmer air:** Arnaud Louis Sountong-Noma Ouedraogo et al., "Comparative Assessment of the Thermal Comfort in Naturally Evolving Buildings in Hot and Dry Sahelian Climate: Case of the City of Ouagadougou in Burkina Faso," *Journal of Materials Science and Surface Engineering* 10 (2023): 1104–11.

165 **Francis Kéré, who hails from the remote village:** Michael Dumiak, "Too Cool for School," *Green Source Magazine of Sustainable Design* (May–June 2010): 65–67.

165 **Kéré began to build:** Diébédo Francis Kéré, "School in Gando, Burkina Faso," *Architectural Design* 82, no. 6 (November 2012): 66–71.

166 **naturally comfortable until late afternoon:** Madlen Kobi, "Thermal Layers: The Case of the Lycée Shorge in Koudougou (Burkina Faso) / Francis Kéré in Conversation with Madlen Kobi," in *The Urban Microclimate as Artifact: Towards an Architectural Theory of Thermal Diversity,* eds. Sascha Roesler and Madlen Kobi (Basel: Birkhauser, 2018), 120–33.

167 **Hyderabad, India, shares a hot and dry climate:** Madhavi Indraganti and Kavita Daryani Rao, "Effect of Age, Gender, Economic Group and Tenure on Thermal Comfort: A Field Study in Residential Buildings in Hot and Dry Climate with Seasonal Variations," *Energy and Buildings* 42 (2010): 273–81.

167 **A study from Ghadames, Libya:** Taki et al., "Assessing Thermal Comfort in Ghadames, Libya."

167 **Burkinabe who live and work:** Èlia Borràs, "'We Don't Need Air Con': How Burkina Faso Builds Schools That Stay Cool in 40C Heat," *The Guardian,* February 29, 2024.

168 **one-fiftieth of 1 percent:** Hannah Ritchie and Max Roser, "Burkina Faso: CO_2 Country Profile," *Our World in Data,* n.d., accessed December 12, 2024, https://ourworldindata.org/co2/country/burkina-faso.

168 **Marrakech and Phoenix:** Aljawabra and Nikolopoulou, "Thermal Comfort in Urban Spaces."

168 **We sweat more:** Allyson Chiu, "Your Body Can Build Up Tolerance to Heat. Here's How," *The Washington Post,* July 29, 2023.

168 **"comfort reparations":** Barber, "After Comfort."

169 **"When it's 140 degrees out":** Elizabeth Royte, "Too Hot to Live: Millions Worldwide Will Face Unbearable Temperatures," *National Geographic,* June 17, 2021.

Chapter 8: A Different Light

170 **work better for women:** Claudia Garcia et al., *Understanding How Women Travel* (Los Angeles County Metropolitan Transportation Authority, August 2019); Kounkuey Design Initiative, *Changing Lanes: A Gender Equity Transportation Study* (Los Angeles Department of Transportation, June 2021).

171 **shared them with the world on Twitter:** On May 18, 2023, Streetsblog LA (@StreetsblogLA) posted photos of the La Sombrita press conference on Twitter at 11:22 A.M., Kounkuey Design Initiative (@Kounkuey) tweeted a photo at 1:56 P.M., and the Los Angeles Department of Transportation (@LADOTofficial) tweeted photos at 5:01 P.M. As of this writing, these posts have been viewed 2 million, 9.5 million, and 833,000 times, respectively.

171 **"unintentionally engineered for internet comedy":** Kriston Capps, "A Defense of 'La Sombrita,' LA's Much-Mocked Bus-Stop Shade," *Bloomberg,* May 25, 2023.

172 **"Well, it doesn't cover anything":** Univision Los Angeles, "'La Sombrita,' El Proyecto Piloto Que Suscita Polémica Entre Usuarios de Autobuses en Los Ángeles," May 22, 2023, YouTube, https://www.youtube.com/watch?v=lD2kRzjwtPc.

172 **"This is ridiculous":** Instagram video posted by De Los (@delosangeles times) on June 14, 2023.

172 **Mayor Karen Bass's public works team scrambled:** Maylin Tu, "LA Mayor's Office Learned About La Sombrita from Press Release," *Los Angeles Public Press,* August 21, 2023.

172 **"If this piece of metal doesn't represent":** TikTok video posted by GoodParty.org (@goodparty) on June 1, 2023.

172 **"Another example of the weird paradox":** Politics & Education (@PoliticsAndEd) posted on Twitter about La Sombrita on May 18, 2023, at 8:10 P.M.

172 **"Performative grift":** @big_pedestrian tweeted about La Sombrita on May 19, 2023, at 2:43 P.M.

172 **"a scam with upbeat PowerPoints":** @PuncheeBurro tweeted on May 18, 2023, at 11:05 P.M.

174 **they've seen *Speed* and watched *Insecure*:** Kendra Pierre-Louis, "Better Bus Systems Could Slow Climate Change," *Scientific American,* May 1, 2023.

174 **around 707,000 passengers:** Estimated ridership statistics for Los Angeles County Metropolitan Transportation Authority (Metro) buses retrieved from https://opa.metro.net/MetroRidership/. Additionally, twenty-eight thousand riders boarded DASH buses operated by the city's transportation department (LADOT) in April 2023, according to Colin Sweeney of LADOT, who corresponded with me in May 2023.

174 **impoverished Latinos who cannot afford a car:** Metro, "2022 Metro Customer Experience Survey," n.d., accessed October 1, 2024, https://www.metro.net/about/survey-results/.

174 **transit agencies spend only 6 percent:** Mary Buchanan and Kirk Hovenkotter, *From Sorry to Superb: Everything You Need to Know About Great Bus Stops* (New York: TransitCenter, October 2018).

174 **sheer number of people:** Ridership statistics in large U.S. and Canadian cities are published quarterly by the American Public Transportation Association and available online.

174 **three-quarters of the city's four thousand or so bus stops:** City of Phoenix, *Shade Phoenix: An Action Plan for Trees and Built Shade,* Draft for City Council Review, November 13, 2024.

174 **compared L.A. bus stops to medieval pillories:** Doug Suisman, "Portals Not Pillories: The Bus Stop and Public Space," *arcCA: Architecture California* 1, no. 3 (2001): 10–13.

175 **half the journey time:** "The Bus Stops Here," *Investing in Place,* September 19, 2022.

175 **Naria Kiani, a former senior planning principal:** Interview on May 19, 2023.

175 **women heat up faster:** Nicola Davis, "Women More at Risk From Heatwaves than Men, Experts Suggest," *The Guardian,* July 18, 2022.

176 **"no way" middle-class people:** Ellen Stern Harris, "Consumer Advocate: Dressing Up the Bus Stop," *Los Angeles Times,* May 4, 1975.

176 **Market research and academic studies**: Philip Law, "The Los Angeles Bus Shelter Program: An Analysis of Location, Design, and Construction Contracts and Recommendations for Improvement" (master's thesis, University of California, Los Angeles, 1999), 9.

176 **accessible to the elderly:** Older adults' most frequent complaint is that bus stops lack shelter or shade, according to a 2020 study by Metro and the Aging and Disability Transportation Network. See Steve Scauzillo, "Most Metro Bus Stops in LA County Have No Shelter, Exposing Riders to Rain, Heat," *Los Angeles Daily News,* February 22, 2023.

176 **shelters have increased ridership:** Ja Young Kim et al., "Another One Rides the Bus? The Connections Between Bus Stop Amenities, Bus Ridership, and ADA Paratransit Demand," *Transportation Research Part A: Policy and Practice* 135 (2020): 280–88.

177 **nearly 1,900 official bus shelters within the city limits:** Maylin Tu, "LA Bus Riders Need Bus Shelters. What They Got Was La Sombrita," *Los Angeles Public Press,* June 1, 2023; Maylin Tu, "More Than 75% of Bus Stops in the City of Los Angeles Have No Shelter. What Now?," *Los Angeles Public Press,* September 26, 2023.

177 **tended to place the shelters:** Madeline Brozen et al., "Are LA Bus Riders Protected from Extreme Heat? Analyzing Bus Shelter Provision in Los Angeles County," UCLA Lewis Center for Regional Policy Studies, Los Angeles, 2023.

177 **politically savvy constituents who didn't ride the bus:** James Rainey,

"Breaking Down Greuel's $160-Million 'Waste, Fraud and Abuse' Savings," *Los Angeles Times,* January 30, 2013; Maylin Tu, "Decades-Long Battle Over Sidewalk Advertising Leaves L.A. Bus Riders Waiting for Shade," PBS SoCal, August 3, 2022.

177 **unhoused person sleeping on their street:** Interview with Lance Oishi and Audrey Netsawang of the City of Los Angeles Department of Public Works, Bureau of Street Services, April 8, 2021.

178 **Kounkuey submitted drawings:** Interview with Jerome Chou, Melissa Guerrero, Christian Rodriguez, and Daniel Garcia of Kounkuey Design Initiative, August 30, 2021. Guerrero and Garcia are no longer with the firm.

178 **there was no public funding:** Lance Oishi, public information session about Sidewalk and Transit Amenities Program (STAP), Zoom, May 19, 2022.

178 **3,400 cantilever umbrellas:** Nam Kyung-don, "3,444 Shade Canopies Installed in Seoul," *The Korea Herald,* June 14, 2024.

178 **more likely to suffer heatstroke:** Eli Lipmen, "Move LA/UCLA Lewis Center Research Reveals Lack of Shelter in LA County," *MoveLA Blog,* October 25, 2023.

178 **"it's a civil rights issue":** Mike Bonin at Los Angeles City Council meeting, September 20, 2022. Bonin left the council later that year.

179 **laborious permitting process:** Alissa Walker, "Here's How Hard Los Angeles Has Made It to Install a Bus Shelter," *Curbed,* July 28, 2021.

179 **width of a commercial sidewalk:** Madeline Brozen et al., *Improving Access to Outdoor Dining Opportunities: Analyzing Constraints of LA Al Fresco* (Los Angeles: UCLA Lewis Center for Regional Policy Studies, 2022).

180 **require at least eight feet of space:** Interview with Lance Oishi and Audrey Netsawang, August 16, 2021.

180 **Since the 1920s:** Martin Wachs et al., *A Century of Fighting Traffic Congestion in Los Angeles, 1920–2020* (Los Angeles: UCLA Luskin Center for History and Policy, 2020).

180 **since the 1960s:** Manville, "Automatic Street Widening."

180 **ask the public works department:** Joe Linton, "City Council Mo-

tion on Ending Automatic Road Widening at Public Works Committee Tomorrow," *Streetsblog LA,* September 24, 2024.

180 **a city with one-third of L.A.'s budget:** In 2024, the city councils of Barcelona and Los Angeles passed budgets of $3.9 billion and $12.8 billion, respectively.

180 **Barcelona has proactively reclaimed:** Interview with Philip Speranza of the University of Oregon, February 3, 2023.

180 **planned to be green:** This history of Barcelona urban design is based on the Speranza interview, an interview with Marc Montlleó of Barcelona Regional on July 28, 2022, an interview with Dani Alsina of Barcelona d'Infraestructures Municipales, SA (BIMSA), and Neda Kostandinovic of the Barcelona City Council on July 28, 2022, and David Roberts, "Barcelona's Remarkable History of Rebirth and Transformation," *Vox,* April 8, 2019.

181 **Europe's highest traffic rate:** Jon Henley et al., "Bollards and 'Superblocks': How Europe's Cities Are Turning on the Car," *The Guardian,* December 18, 2023.

182 **Roads comprise about 60 percent:** Manel Ferri Tomàs, "Superar la vieja movilidad. El coche, de anfitrión a invitado," *Barcelona Metròpolis* 117 (2020).

182 **a major program of traffic calming:** Alsina and Kostandinovic interview.

182 **A plan for a citywide network:** David Roberts, "Barcelona Wants to Build 500 Superblocks. Here's What It Learned from the First Ones," *Vox,* April 9, 2019.

182 **one unified department:** Elisa Muzzini and Sophia Torres, *Thematic Review: Resilience in Action: Barcelona's Superblock Programme* (Paris: Council of Europe Development Bank, September 2023).

183 **Let's fill the streets with life!:** Comissió d'Ecologia, Urbanisme i Mobilitat, *Omplim de vida els carrers: La implantació de les superilles a Barcelona* (Barcelona: City Council, May 2016).

183 **new kind of urban green:** Alsina and Kostandinovic interview.

184 **achieved an ultraefficient system:** Interviews with Gabino Carballo of the Barcelona Municipal Institute of Parks and Gardens, July 29,

2022, and Gerard Marias of Barcelona Cicle de l'Aigua, SA (BCASA), and Dani Alsina, July 29, 2022.

184 **an urban forest of that magnitude:** Tamara Iungman et al., "Cooling Cities Through Urban Green Infrastructure: A Health Impact Assessment of European Cities," *The Lancet* 401, no. 10376 (February 2023): 577–89.

184 **additionally removing cars:** Natalie Mueller et al., "Changing the Urban Design of Cities for Health: The Superblock Model," *Environment International* 134 (January 2020): 105132.

185 **billions in healthcare costs:** Natalie Mueller et al., "Urban and Transport Planning Related Exposures and Mortality: A Health Impact Assessment for Cities," *Environmental Health Perspectives* 125, no. 1 (January 2017): 89–96.

185 **copying Barcelona's superblocks:** Iain Campbell et al., *Beating the Heat: A Sustainable Cooling Handbook for Cities* (Nairobi: United Nations Environment Programme, 2021), 97.

185 **60,000 parking spaces:** Feargus O'Sullivan, "Paris to Replace Parking Spaces with Trees," *Bloomberg CityLab,* November 18, 2024.

185 **reclaiming car habitat for trees:** Interview with Cecil Konijnendijk of the Nature Based Solutions Institute, February 17, 2023.

186 **"What is wrong with all you people?":** Interview with Ariane Middel, May 20, 2023.

186 **perfectly acceptable substitutes for trees:** Ariane Middel et al., "50 Grades of Shade," *Bulletin of the American Meteorological Society* 102, no. 9 (May 2021): 1–35.

186 **"shade deserts":** V. Kelly Turner et al., "Shade Is an Essential Solution for Hotter Cities," *Nature* 619 (2023): 694–97.

187 **"five-foot-ways":** Ken Yeang, *The Tropical Verandah City: Some Urban Design Ideas for Kuala Lumpur* (Petaling Jaya: Longman Malaysia, 1987), 35–36.

187 **interest in climate and comfort:** Cherian George, *The Air-Conditioned Nation: Essays on the Politics of Comfort and Control, 1990–2000* (Singapore: Landmark Books, 2000), 13–14.

187 **communal "void decks":** Interview with Jiat-Hwee Chang of the National University of Singapore, November 2, 2021.

188 **125 miles of covered walkways:** Rachel Au-Yong, "Walk2Ride

Scheme to Extend Walkways Will Hit 200 km Milestone on Sept 19," *The Straits Times,* September 15, 2018.

188 **twelve feet of pedestrian overhangs:** Singapore Urban Redevelopment Authority (URA), Development Control, *Commercial Handbook: Provision of Open and Covered Walkways,* n.d., accessed October 1, 2024, https://www.ura.gov.sg/Corporate/Guidelines/Development-Control/Non-Residential/Commercial/Covered-Walkways.

188 **feels 14 percent shorter:** Valentin R. Melnikov et al., "Behavioural Thermal Regulation Explains Pedestrian Path Choices in Hot Urban Environments," *Scientific Reports* 12 (2022): 2441.

188 **moisture can add to the misery:** Interview with Lea Ruefenacht, May 28, 2021.

188 **authorities consider "sufficient" shade:** URA, *Update to the Design Guidelines for Privately Owned Public Spaces (POPS)* (Circular URA/PB/2022/07-AUDG), Annex 3, June 3, 2022.

189 **not allowed to face north:** New York City Department of City Planning, "New York City's Privately Owned Public Spaces, Current Standards, Restrictions on Orientation," n.d., accessed October 1, 2024, https://www.nyc.gov/site/planning/plans/pops/pops-plaza-standards.page.

189 **plazas on the east side:** Nicholas Li, *A Green and Liveable City: Singapore Urban Design Guidebook* (Singapore: URA, 2023), 124–25.

189 **"planning codes had to encourage shade":** Interview with Kelvin Ang, November 10, 2021. For critiques of urban design in Israel and Sri Lanka, see David Pearlmutter, "Patterns of Sustainability in Desert Architecture," *Arid Lands Newsletter* 47 (May 2000); Rohinton Emmanuel, ed., *Urban Climate Challenges in the Tropics* (London: Imperial College Press, 2016), 5–7.

189 **"give me shade first":** Ng Tze Yong, "'Flowers? Give Me Shade First,'" *The New Paper,* May 18, 2009.

189 **Lee ordered his land use:** Richard Webb, "Urban Forestry in Singapore," *Arboricultural Journal* 22, no. 3 (1998): 271–86.

190 **budget increased tenfold:** Belinda Yuen, "Creating the Garden City: The Singapore Experience," *Urban Studies* 33, no. 6 (June 1996): 955–70.

190 **"We did not differentiate":** Lee Kuan Yew, *From Third World to First: The Singapore Story: 1965–2000* (New York: HarperCollins, 2000), 175–76.

190 **urban forest grew from 158,600 trees:** Lim Tin Seng, "Of Parks, Gardens and Trees: The Greening of Singapore," *BiblioAsia* 17, no. 1 (April–June 2021): 62–68.

190 **"this is one good thing":** Interview with Daniel Burcham, November 29, 2021.

191 **streets of Singapore's business district:** Marta Chàfer et al., "Mobile Measurements of Microclimatic Variables Through the Central Area of Singapore: An Analysis from the Pedestrian Perspective," *Sustainable Cities and Society* 83 (August 2022): 103986.

191 **green grounds of a public housing estate:** Winston Chow and Matthias Roth, "Temporal Dynamics of the Urban Heat Island of Singapore," *International Journal of Climatology* 26, no. 15 (July 2006): 2243–60.

192 **skin cancer diagnoses:** Sax Institute, "Two in Three Australians Will Develop Common Skin Cancers as Incidence Rises, New Research Shows," press release, March 10, 2022.

193 **rate has more than doubled:** Cancer Council, "Skin Cancer Incidence and Mortality," n.d., accessed October 20, 2024, https://www.cancer.org.au/about-us/policy-and-advocacy/prevention/uv-radiation/related-resources/skin-cancer-incidence-and-mortality.

193 **"Slip! Slop! Slap!":** Robin Marks, "From Slip! Slop! Slap! to SunSmart—the Public Health Approach to Skin Cancer Control in Australia in the 1990s," in *The Environmental Threat to the Skin,* ed. Ronald Marks and Gerd Plewig (London: Martin Dunitz, 1992), 39–42.

193 **"Habitual behavior starts in childhood":** David Hill, "The Birth of Slip, Slop, Slap and the Science Behind Nation-Wide Behaviour Change," *Cancer Council Blog,* Cancer Council, January 23, 2020, https://www.cancer.org.au/blog/the-birth-of-slip-slop-slap-and-the-science-behind-nation-wide-behaviour-change.

194 **playgrounds in wealthier neighborhoods:** Caroline Anderson et al., "Shade in Urban Playgrounds in Sydney and Inequities in Availability

for Those Living in Lower Socioeconomic Areas," *Australian and New Zealand Journal of Public Health* 38, no. 1 (February 2014): 49–53.

194 **"somebody has to make sure":** Interview with John Greenwood, March 22, 2020.

194 **At a Sydney elementary school:** John Greenwood et al., *Under Cover: Guidelines for Shade Planning and Design* (Sydney: New South Wales Health Department, 1998): 177–85.

195 **sun protection for adolescents:** Craig Sinclair and Peter Foley, "Skin Cancer Prevention in Australia," *British Journal of Dermatology* 161, no. S3 (November 2009): 116-23.

195 **"natural and built shade":** Greenwood, *Under Cover,* 53.

196 **run their own modified audits:** The Cancer Council of Victoria offers a "shade comparison check" online tool at https://www.sunsmart.com.au/resources/shade-comparison-check (n.d., accessed October 20, 2024).

196 **New South Wales public playgrounds:** Cancer Institute NSW, *Summary Report of Benchmarking Shade in NSW Playgrounds* (Sydney: Cancer Institute NSW, 2023).

196 **primary schools in Victoria:** Barbara Chancellor, "Primary School Playgrounds: Features and Management in Victoria, Australia," *International Journal of Play* 2, no. 2 (2013): 63–75.

196 **only 33 percent of American playgrounds:** Eric Kennedy et al., *Thermally Comfortable Playgrounds: A Review of Literature and Survey of Experts* (Cedar Falls, Iowa: National Program for Playground Safety at the University of Northern Iowa, 2020).

196 **Population surveys in Melbourne:** Tamara Tabbakh et al., "Implementation of the SunSmart Program and Population Sun Protection Behaviour in Melbourne, Australia: Results from Cross-Sectional Summer Surveys from 1987 to 2017," *PLoS Medicine* 16, no. 10 (2019): e1002932.

196 **people under thirty years old:** Anne E. Cust et al., "What Is Behind the Declining Incidence of Melanoma in Younger Australians?," *The Medical Journal of Australia* 221, no. 5 (2024): 246–47.

196 **"It's kind of like saying":** Interview with David Buller, January 10,

2023. For study results, see Buller et al., "Shade Sails and Passive Recreation in Public Parks of Melbourne and Denver: A Randomized Intervention," *American Journal of Public Health* 107, no. 12 (December 2017): 1869–75.

196 **closing the ozone hole:** Terry Slevin and David Whiteman, "Why Does Australia Have So Much Skin Cancer? (Hint: It's Not Because of an Ozone Hole)," *The Conversation,* March 19, 2018.

198 **"In talking to constituents":** Amanda Shendruk, "How to Transform City Streets—Without Losing Your Parking Spot," *The Washington Post,* August 13, 2024.

198 **mobility of older adults:** Anastasia Loukaitou-Sideris, "The Right to Walk in the Neighborhood: Designing Inclusive Sidewalks for Older Adults," in *Just Urban Design: The Struggle for a Public City,* eds. Kian Goh et al. (Cambridge, Mass.: MIT Press, 2022), 295–314.

199 **children feel better:** Jennifer Vanos et al., "Effects of Physical Activity and Shade on the Heat Balance and Thermal Perceptions of Children in a Playground Microclimate," *Building and Environment* 126 (2017): 119–31.

199 **few American cities consider the expansion:** V. Kelly Turner et al., "How Are Cities Planning for Heat? Analysis of United States Municipal Plans," *Environmental Research Letters* 17, no. 6 (2022): 064054.

Chapter 9: Making Sunsets

202 **"a Burning Man camp with fast internet and electricity":** Interviews with Luke Iseman and Andrew Song, December 2022–February 2023.

203 **Sulfur dioxide is nasty stuff:** Alejandro de la Garza, "Exclusive: Inside a Controversial Startup's Risky Attempt to Control Our Climate," *Time,* February 21, 2023.

206 **change the temperature of the earth:** This assumes low-albedo shading devices.

206 **Trees offer a notable exception:** Evaporative cooling does not directly affect the global energy balance. See George A. Ban-Weiss et al., "Climate Forcing and Response to Idealized Changes in Surface La-

tent and Sensible Heat," *Environmental Research Letters* 6, no. 3 (2011): 034032.

206 **block less than 2 percent of the light:** Jocelyn Kaiser, "A Sunshade for Planet Earth," *Science,* October 31, 2006; Joe Palca, "Telescope Innovator Shines His Genius on New Fields," NPR, August 23, 2012.

207 **A giant solar sail:** According to the Planetary Sunshade Foundation, the sunshade would block one-half of 1 percent of incoming solar radiation and cool the planet by 1 degree Celsius. See *State of Space-Based Solar Radiation Management* (Los Angeles: Planetary Sunshade Foundation, March 2023).

207 **"Skeptics might tell you there is no way":** Mark Matthews, "Dimmer Switch," *ASEE Prism* 31, no. 1 (Fall 2022): 22–27.

207 **a close look at planetary shade:** Corbin Hiar, "Blocking Sun Rays Finds Support in the Senate," *E&E News by Politico,* March 2, 2023.

208 **restore a preindustrial climate:** David Keith and Peter Irvine, "Solar Geoengineering Could Substantially Reduce Climate Risks—a Research Hypothesis for the Next Decade," *Earth's Future* 4, no. 11 (2016): 549–59.

208 **"There are many people":** Interview with Janos Pasztor, October 2, 2023.

209 **the window has practically closed:** The IPCC's Sixth Assessment Report predicted Earth would cross 1.5 degrees Celsius of warming by the early 2030s; the threshold was exceeded in 2024.

209 **2.6 to 3.1 degrees Celsius:** United Nations Environment Programme, *Emissions Gap Report 2024: No More Hot Air . . . Please!* (Nairobi: United Nations Environment Programme, 2024).

209 **"dangerous fantasy":** David Keith, "What's the Least Bad Way to Cool the Planet?," *The New York Times,* October 1, 2021.

210 **bounce 30 to 70 percent of the sunlight:** Natalie Wolchover, "A World Without Clouds," *Quanta Magazine,* February 25, 2019.

210 **3 percent increase in reflectivity:** Christopher Flavelle, "Warming Is Getting Worse. So They Just Tested a Way to Deflect the Sun," *The New York Times,* April 2, 2024.

210 **"albedo yachts":** Christopher Mims, "'Albedo Yachts' and Marine

Clouds: A Cure for Climate Change?," *Scientific American,* October 21, 2009.

210 **fog the air above the beloved corals:** Graham Readfearn, "Scientists Trial Cloud Brightening Equipment to Shade and Cool Great Barrier Reef," *The Guardian,* April 16, 2020; Jeff Tollefson, "Can Artificially Altered Clouds Save the Great Barrier Reef?," *Nature* 596 (August 2021): 476–78.

211 **delayed the bleaching:** Peter Butcherine et al., "Intermittent Shading Can Moderate Coral Bleaching on Shallow Reefs," *Frontiers in Marine Science* 10 (September 2023).

211 **moved quickly to shut it down:** Anthony Edwards, "City of Alameda Stops Controversial Geoengineering Experiment," *San Francisco Chronicle,* May 13, 2024.

211 **Australian scientists encountered less resistance:** Jesse Reynolds and Pete Irvine, "29. Daniel Harrison on Marine Cloud Brightening and the RRAP," *Challenging Climate* (podcast), February 7, 2023.

212 **Russian climatologist named Mikhail Budyko:** A. Lapenis, "A 50-Year-Old Global Warming Forecast That Still Holds Up," *Eos,* November 25, 2020.

212 **Nobel Prize–winning chemist Paul Crutzen:** Gernot Wagner, *Geoengineering: The Gamble* (Cambridge, UK: Polity Press, 2021), 3–4.

212 **scientists like James Hansen:** Seth Borenstein, "Pioneering Scientist Says Global Warming Is Accelerating. Some Experts Call His Claims Overheated," Associated Press, November 2, 2023.

213 **this was a "pious wish":** Paul Crutzen, "Albedo Enhancement by Stratospheric Sulfur Injections: A Contribution to Resolve a Policy Dilemma?," *Climatic Change* 77 (2006): 211–20.

213 **a million tons of smoke into the stratosphere:** Madeleine Stone, "How Extreme Fire Weather Can Cool the Planet," *National Geographic,* August 6, 2021.

214 **silver lining of Smokemageddon:** Matthew Zeitlin, "The Smoke Is Actually Making Us Colder—for Now," *Heatmap,* July 1, 2023; Peter Coy, "The Only Positive of Smokemageddon," *The New York Times,* June 9, 2023.

214 **best method of particle injection:** Wake Smith and Gernot Wagner,

"Stratospheric Aerosol Injection Tactics and Costs in the First 15 Years of Deployment," *Environmental Research Letters* 13 (2018): 124001; Wagner, *Gamble,* 18–20.

214 **one-tenth of a degree:** Jesse L. Reynolds and Gernot Wagner, "Highly Decentralized Solar Geoengineering," *Environmental Politics* 29, no. 5 (2020): 917–33.

215 **devastating impact on Africa:** Daisy Dunne, "Unregulated Solar Geoengineering Could Spark Droughts and Hurricanes, Study Warns," *Carbon Brief,* November 14, 2017.

215 **"we let them eat the risk":** Pablo Suarez and Maarten K. van Aalst, "Geoengineering: A Humanitarian Concern," *Earth's Future* 5, no. 2 (February 2017): 183–95.

215 **backed a "non-use agreement":** Solar Geoengineering Non-Use Agreement, n.d., accessed December 12, 2024, https://www.solar geoeng.org.

216 **less than sixty seconds' worth of condensation trails:** Wagner, *Gamble,* 81.

216 **pushback from the Sámi:** Alister Doyle, "Indigenous Peoples Urge Harvard to Scrap Solar Geoengineering Project," Reuters, June 9, 2021.

216 **reindeer are starving:** Environmental Justice Foundation, *Rights at Risk: Arctic Climate Change and the Threat to Sami Culture* (London: EJF, June 2019).

216 **"overbelief of technology":** Mark Turner, "C2GTalk: Why Did the Saami Council Oppose Harvard's SCoPEx Experiment? With Åsa Larsson Blind," *Carnegie Council Podcasts* (podcast), December 12, 2022.

217 **"not put airbags in cars":** Paul Rand, "A Radical Solution to Address Climate Change, with David Keith," *Big Brains* (podcast), November 30, 2023.

217 **"we won't be open":** David Gelles, "This Scientist Has a Risky Plan to Cool Earth. There's Growing Interest," *The New York Times,* August 1, 2024.

217 **"Private philanthropy money":** Gernot Wagner, "Can Geoengineering Slow Climate Change? We Need Research to Find Out," *The Washington Post,* February 22, 2023.

217 **coterie of online enthusiasts:** For example, Silicon Valley executive Tomás Pueyo endorsed Make Sunsets on Substack ("How You Can Easily Delay Climate Change Today: SO2 Injection," *Uncharted Territories,* March 5, 2024), and popular tweeter Crémieux (@cremieuxrecueil) called Make Sunsets "the solution" to warming caused by clean air in a May 25, 2024, thread on X.

217 **Iseman has told the world:** David Gelles, "Renegades of Silicon Valley Pollute the Sky to Save the Planet," *The New York Times,* September 30, 2024; Ben Tracy, "To Mitigate Impacts of Climate Change, Some Turn to Controversial 'Geoengineering,'" *CBS Saturday Morning,* April 22, 2023; Julia Simon, "The Sunday Story: Startups Want to Cool Earth by Reflecting Sunlight," *Up First from NPR,* April 28, 2024.

218 **two venture capital firms to fund him:** James Temple, "A Startup Says It's Begun Releasing Particles into the Atmosphere, in an Effort to Tweak the Climate," *MIT Technology Review,* December 24, 2022.

218 **island nation of Kiribati:** Make Sunsets, email to mailing list subscribers, December 31, 2024.

218 **financial motives to oversell the benefits:** Temple, "Startup Says It's Begun Releasing Particles."

218 **"I'm going to speak bluntly":** Jesse Reynolds and Peter Irvine, "27. Luke Iseman on His For-Profit Solar Geoengineering Venture—Make Sunsets," *Challenging Climate* (podcast), January 10, 2023.

219 **tech giants are among the biggest investors:** Corbin Hiar, "Inside EDF's Private Meeting on Geoengineering," *E&E News by Politico,* February 14, 2024; Stephen Robert Miller, "Inside Silicon Valley's Grand Ambitions to Control Our Planet's Thermostat," *Noema Magazine,* August 8, 2024.

219 **"unfuck the planet":** Tim Crino, "How VC Chris Sacca Plans to Make Trillions on Climate Change," *Inc.,* April 29, 2022.

219 **"it is anti-democratic":** Shuchi Talati (@sktalati) post on X, May 25, 2024, 12:53 P.M.

219 **lives lost to climate change:** R. Daniel Bressler, "The Mortality Cost of Carbon," *Nature Communications* 12, no. 1 (2021): 4467.

220 **policymakers contemplate paths:** White House Office of Science

and Technology Policy, *Congressionally Mandated Research Plan and an Initial Research Governance Framework Related to Solar Radiation Modification* (Washington, D.C.: OSTP, 2023); Christopher Flavelle and David Gelles, "U.K. to Fund 'Small-Scale' Outdoor Geoengineering Tests," *The New York Times,* September 13, 2024.

220 **funded by Western charities and donors:** Anja Chalmin, "Global Southwashing: How The Degrees Initiative Is Imposing Its Solar Geoengineering Agenda onto Climate Research in the Global South," *Geoengineering Monitor,* October 16, 2024.

220 **probe the impact on their countries:** James Temple, "This Technology Could Alter the Entire Planet. These Groups Want Every Nation to Have a Say," *MIT Technology Review,* April 17, 2023.

220 **governance could be led:** Interviews with Shuchi Talati on April 3, 2023, and Talati and Hassaan Sipra on September 18, 2024.

221 **implications of climate change:** Hassaan Sipra, "Moving Beyond Polarization on Climate Interventions," panel at Climate Week NYC, New York, September 25, 2024.

221 **In the case of Pakistan:** Imran Saqib Khalid and Hassaan Sipra, *Governance of Solar Radiation Modification: Developing the Pakistan Perspective* (Washington, D.C.: Alliance for Just Deliberation on Solar Geoengineering, February 2024).

221 **"When to stop?":** "Moving Beyond" panel.

221 **entrepreneurs like Bill Gates:** The Microsoft founder was an early funder of Keith's geoengineering research. See Wagner, *Gamble,* 24.

222 **"it's now or never":** Talati interview, September 2024.

222 **Montreal Protocol to ban the chemicals:** United Nations Environment Programme, "Ozone Layer Recovery Is on Track, Helping Avoid Global Warming by 0.5°C," press release, January 9, 2023.

223 **met with hundreds of government officials:** Janos Pasztor, "We're Likely to Overshoot the Paris Goals—and We Need to Start Talking About It," *Politico.eu,* September 21, 2022.

224 **"Wait until Mr. Musk does this":** Pasztor interview.

224 **Even this small step:** Duncan McLaren and Olaf Corry, "Countries Failed to Agree First Steps on Solar Geoengineering at the UN. What Went Wrong?," *Legal Planet,* March 7, 2024.

224 **countries could not come together:** Pasztor (@jpasztor) post on X, February 29, 2024, 11:26 A.M.

225 **"Stardust exists":** Pasztor (@jpasztor) post on X, May 4, 2024, 3:37 P.M.

226 **"So fucking amateur":** De la Garza, "Inside a Controversial Startup's Risky Attempt."

Art Credits

page 26: *Español: Vista de la calle Sierpes de Sevilla en 1918* (Spanish: View of Sierpes Street in Seville in 1918). Anonymous/Alamy.

page 56: Simplified version of the bioclimatic chart by Victor and Aladar Olgyay. Adapted from Housing and Home Finance Agency, *Application of Climatic Data to House Design* (Washington, D.C.: U.S. Government Printing Office, 1954), 20.

page 59: Scruggs, Vandervoort and Barney department store in Clayton, Missouri, by architect Harris Armstrong. Photograph by Julius Shulman, 1952. © J. Paul Getty Trust. Getty Research Institute, Los Angeles (2004.R.10).

page 154: Skidmore *Passivhaus* (Passive House) in Portland, Oregon, by architect Jeff Stern. Photograph by Jeremy Bitterman/JBSA, 2013.

page 166: Gando Primary School by architect Francis Kéré in Burkina Faso. Photograph by Erik-Jan Ouwerkerk.

page 173: La Sombrita shade structure on Saticoy Street in Los Angeles. Photograph by Adam Baron-Bloch, 2024.

Index

About the Author

Sam Bloch is an environmental journalist. Previously a staff writer at *The Counter,* he has written for *LA Weekly, Places Journal, Slate, The New York Times, CityLab,* and *Landscape Architecture Magazine,* among others. Bloch is a graduate of the Columbia Journalism School and a former MIT Knight Science Journalism Fellow and Emerson Collective Fellow. He is based in New York City.

About the Type

This book was set in Bembo, a typeface based on an old-style Roman face that was used for Cardinal Pietro Bembo's tract *De Aetna* in 1495. Bembo was cut by Francesco Griffo (1450–1518) in the early sixteenth century for Italian Renaissance printer and publisher Aldus Manutius (1449–1515). The Lanston Monotype Company of Philadelphia brought the well-proportioned letterforms of Bembo to the United States in the 1930s.